Collins

SECURE SCIENCE FOR GCSE

Workbook

INTERVENTION TO GET BACK ON TRACK

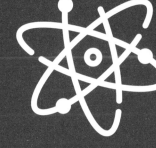

Lucy Wood

William Collins' dream of knowledge for all began with the publication of his first book in 1819.
A self-educated mill worker, he not only enriched millions of lives, but also founded a flourishing publishing house.
Today, staying true to this spirit, Collins books are packed with inspiration, innovation and practical expertise.
They place you at the centre of a world of possibility and give you exactly what you need to explore it.

Collins. Freedom to teach.

Published by Collins
An imprint of HarperCollins*Publishers*
The News Building, 1 London Bridge Street, London, SE1 9GF, UK

HarperCollins*Publishers*
1st Floor, Watermarque Building, Ringsend Road, Dublin 4, Ireland

> **Browse the complete Collins catalogue at
> www.collins.co.uk**

10 9 8 7 6 5 4 3 2 1

ISBN 978-0-00-849209-0

British Library Cataloguing-in-Publication Data
A catalogue record for this publication is available from the British Library.

Authors: Lucy Wood, Jeremy Pollard, Sarah Jinks, Dorothy Warren
Publisher: Katie Sergeant
Product manager: Joanna Ramsay
Product developer: Holly Woolnough
Development editor: Rebecca Ramsden
Copyeditor: Jan Schubert
Proofreader: Helius
Cover designer: Kneath Associates
Cover illustrations: Mialapi/Shutterstock
Illustrations: Ann Paganuzzi
Typesetter: Ken Vail Graphic Design
Production controller: Katharine Willard
Printed and bound by: Bell and Bain Ltd, Glasgow

MIX
Paper from
responsible sources
FSC www.fsc.org **FSC™ C007454**

This book is produced from independently certified FSC™ paper
to ensure responsible forest management.

For more information visit: **www.harpercollins.co.uk/green**

Acknowledgements
The publishers gratefully acknowledge the permission granted to reproduce the copyright material in this book. Every effort has been made to trace copyright holders and to obtain their permission for the use of copyright material. The publishers will gladly receive any information enabling them to rectify any error or omission at the first opportunity.

Contents

How to use this book

Welcome to Secure Science for GCSE!

We hope that this resource will help you to build and secure a deep understanding of GCSE Science, to fix gaps, make connections and solve science problems successfully. There are 30 sessions covering all three sciences including maths, practical and working scientifically skills.

Here's how it works:

THINK

You have been learning science for a really long time so the chances are you will know something about every possible exam question. The first step in each *Secure Science* session helps you remember what you already know. This is called retrieval, and the more you do it, the more you will remember. Each session has a writing frame so you can organise your thoughts. Look at the session question and scribble down what you know on that topic.

Sometimes the information you need can be buried pretty deep, so we have included a deck of cards in the teacher resource with snippets of information that will help you remember things you learned some time ago. Ask your teachers for the cards so you can deal out the relevant cards and use them to add to what you could remember. This will help to get your mind working and thinking about the topic to activate and strengthen your memory. Don't worry if you've got any blanks. Then you can have a go at the questions for the session on Adapt©. Your teacher or tutor can give you access to this platform.

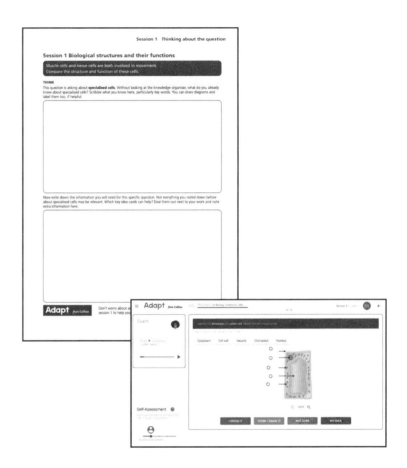

CONNECT

Watch the video on Adapt© about the powerful idea underpinning this topic. Your teacher or tutor can give you access to Adapt©. The knowledge organisers in *Secure Science for GCSE* show how all the ideas link together. To answer the exam-style questions you can fill in any gaps in what you can remember by looking at the knowledge organiser which summarises the key information and how it links together. This will help you make connections and understand underlying principles. This means that if you complete all the sessions you will know something about any exam questions that you could possibly be asked.

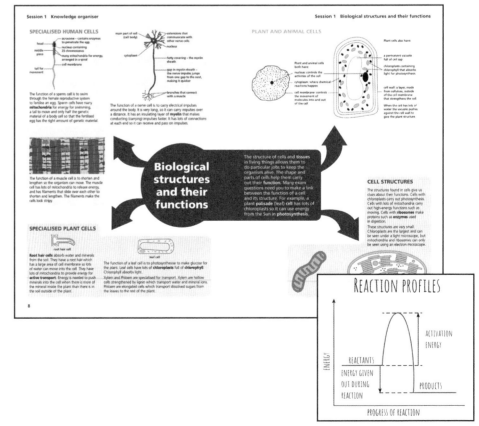

SOLVE

Once you have organised all your ideas you can start to solve the session question. The writing frame will walk you through the question. The frame will help you use the right vocabulary, work out which ideas are relevant and how to structure your answer. Once you've completed the session, turn to the contents page and tick the box to show that it's complete.

PRACTISE

Practice really does make perfect! Students who practice answering questions on different topics over time tend to do better in exams. Repeat the sessions on Adapt© where you want to check and strengthen your knowledge. After the sessions is a section packed full of practice questions that cover each type of exam question requiring a longer answer and the topics from the sessions. You could even use the writing frames to practice answering exam questions from past papers – just find the frame for the command word in your paper. If you get stuck, you'll find the information you need to get unstuck in the knowledge organisers.

I hope you enjoy revising science using *Secure Science for GCSE*, and that you go into your exams feeling confident that you know something about all the topics on the paper. If you do all the *Secure Science* sessions, you will have answered the different types of questions you can be asked and made sense of the key ideas you'll need to make sense of the science in the exams and in your life.

Good luck!

Introduction

Welcome from author Lucy Wood

The book you are holding is designed to help you learn science quickly and effectively: in just 30 sessions, you will develop an understanding of science ideas that will help you feel confident going into the exam and understand the science you will come across in your life outside school. I have used everything I learned as a science teacher and head of science to write this, and then checked with the best science teachers I know from around the country to make sure *Secure Science for GCSE* is the perfect learning tool.

I hope your science learning so far has been a brilliant experience, and you have been able to do lots of interesting work with teachers who have made it make sense. Don't worry if it feels like you have a huge amount to learn right now. GCSE science is enormous: that is why you get two GCSEs. However, *Secure Science* distils (pun intended) all that content down into 30 sessions, and helps you think through exam questions. The *Secure Science* system mimics how the best science teachers I know teach their students to get great exam results and uses research about how we all learn.

Session 1 Biological structures and their functions

Muscle cells and nerve cells are both involved in movement.
Compare the structure and function of these cells.

THINK

This question is asking about **specialised cells**. Without looking at the knowledge organiser, what do you already know about specialised cells? Scribble what you know here, particularly key words. You can draw diagrams and label them too, if helpful.

Now write down the information you will need for this specific question. Not everything you noted down before about specialised cells may be relevant. Which key idea cards can help? Deal them out next to your work and note extra information here.

Adapt *from* Collins   Don't worry about any blanks. Now have a go at the questions on Adapt© for session 1 to help you think further about this topic.

SPECIALISED HUMAN CELLS

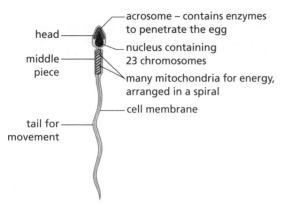

head —
middle piece —
acrosome – contains enzymes to penetrate the egg
nucleus containing 23 chromosomes
many mitochondria for energy, arranged in a spiral
cell membrane
tail for movement

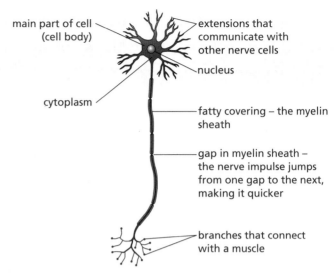

main part of cell (cell body)
cytoplasm
extensions that communicate with other nerve cells
nucleus
fatty covering – the myelin sheath
gap in myelin sheath – the nerve impulse jumps from one gap to the next, making it quicker
branches that connect with a muscle

The function of a sperm cell is to swim through the female reproductive system to fertilise an egg. Sperm cells have many **mitochondria** for energy for swimming, a tail to move and only half the genetic material of a body cell so that the fertilised egg has the right amount of genetic material.

The function of a nerve cell is to carry electrical impulses around the body. It is very long, so it can carry impulses over a distance. It has an insulating layer of **myelin** that makes conducting (carrying) impulses faster. It has lots of connections at each end so it can receive and pass on impulses.

The function of a muscle cell is to shorten and lengthen so the organism can move. The muscle cell has lots of mitochondria to release energy, and has filaments that slide over each other to shorten and lengthen. The filaments make the cells look stripy.

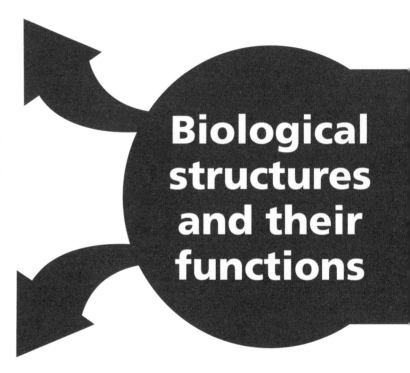

Biological structures and their functions

SPECIALISED PLANT CELLS

root hair cell

Root hair cells absorb water and minerals from the soil. They have a root hair which has a large area of cell membrane so lots of water can move into the cell. They have lots of mitochondria to provide energy for **active transport**. Energy is needed to push minerals into the cell when there is more of the mineral inside the plant than there is in the soil outside of the plant.

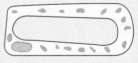

leaf cell

The function of a leaf cell is to photosynthesise to make glucose for the plant. Leaf cells have lots of **chloroplasts** full of **chlorophyll**. Chlorophyll absorbs light.

Xylem and Phloem are specialised for transport. Xylem are hollow cells strengthened by lignin which transport water and mineral ions. Phloem are elongated cells which transport dissolved sugars from the leaves to the rest of the plant.

PLANT AND ANIMAL CELLS

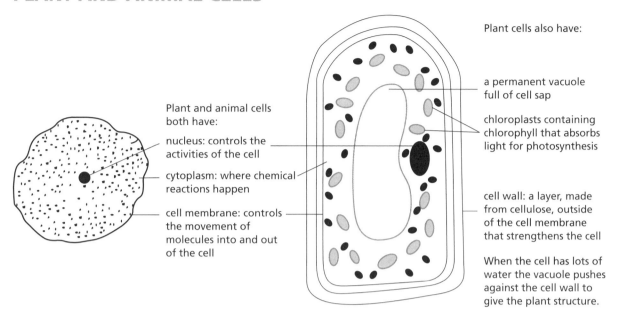

Plant and animal cells both have:

nucleus: controls the activities of the cell

cytoplasm: where chemical reactions happen

cell membrane: controls the movement of molecules into and out of the cell

Plant cells also have:

a permanent vacuole full of cell sap

chloroplasts containing chlorophyll that absorbs light for photosynthesis

cell wall: a layer, made from cellulose, outside of the cell membrane that strengthens the cell

When the cell has lots of water the vacuole pushes against the cell wall to give the plant structure.

The structure of cells and **tissues** in living things allows them to do particular jobs to keep the organism alive. The shape and parts of cells help them carry out their **function**. Many exam questions need you to make a link between the function of a cell and its structure. For example, a plant **palisade** (leaf) **cell** has lots of chloroplasts so it can use energy from the Sun in **photosynthesis**.

CELL STRUCTURES

The structures found in cells give us clues about their functions. Cells with chloroplasts carry out photosynthesis. Cells with lots of mitochondria carry out high-energy functions such as moving. Cells with **ribosomes** make proteins such as **enzymes** used in digestion.

These structures are very small. Chloroplasts are the largest and can be seen under a light microscope, but mitochondria and ribosomes can only be seen using an electron microscope.

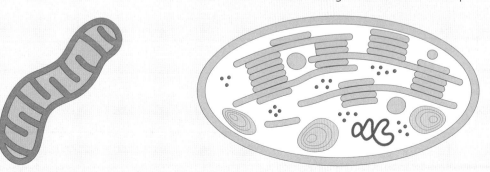

Session 1   Solving the question

> Muscle cells and nerve cells are both involved in movement.
> Compare the structure and function of these cells.

SOLVE

Now you have gathered the relevant information it is really important that you think carefully about how you need to use and present that information. Use the prompts below to help you.

This question asks you to **compare**. This means that you need to talk about similarities and differences. Which structures do muscle cells and nerve cells have in common?

How is the shape of each cell different?

Which structures are different in muscle cells and nerve cells?

For each of the differences, explain how that structure or shape helps the cell do its job.

Now bring together your answers: What do both cells have that they use to make movement happen? What does each cell have that is different that helps it to do its specific job?

There are additional practice questions with the writing frames for 'Calculate', 'Explain' and 'Suggest' at the end of the book.

Session 2 Moving molecules in living things

Plants need water for important reactions like photosynthesis.
Explain why plants wilt when they are not watered.

THINK

This question is asking about the role of water in plants. Plants need water to stand up. Without looking at the knowledge organiser, what do you already know about the structure of plant cells? What do you know about what water does and how it moves through a plant? Scribble what you know here, particularly key words.

Now write down the information you will need for this specific question. Not everything you noted down before about the role of water in plants may be relevant.

Which key idea cards can help? Deal them out next to your work and note extra information here.

Adapt *from Collins*

Don't worry about any blanks. Now have a go at the questions on Adapt© for session 2 to help you think further about this topic.

DIFFUSION

high concentration

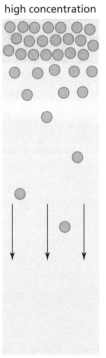

low concentration

Diffusion is the movement of particles from an area of high concentration to an area of low concentration along a concentration gradient.

OSMOSIS

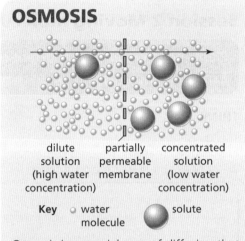

| dilute solution (high water concentration) | partially permeable membrane | concentrated solution (low water concentration) |

Key ○ water molecule ● solute

Osmosis is a special case of diffusion that happens in living things. A dilute solution contains many water molecules (a high concentration of water molecules). A concentrated solution contains fewer water molecules. Osmosis is the movement of water molecules from a dilute to a concentrated solution, across a **partially permeable membrane**.

Moving molecules in living things

MOVING WATER THROUGH A PLANT

water enters the cells in the leaf

water pulled up the xylem

water evaporates from the stomata

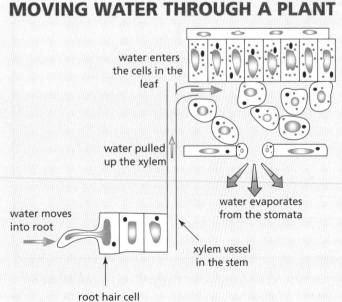

water moves into root

xylem vessel in the stem

root hair cell

Water moves through a plant in different ways. The leaves use up water in photosynthesis, so there is a low concentration of water inside the plant compared with the soil. Water moves into the root by osmosis and up through the stem towards the leaves in a process called **transpiration**. The water is pulled because water is leaving the leaf by evaporation.

DIFFUSION IN PLANTS

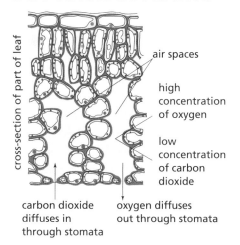

cross-section of part of leaf

air spaces

high concentration of oxygen

low concentration of carbon dioxide

carbon dioxide diffuses in through stomata

oxygen diffuses out through stomata

OSMOSIS IN ANIMAL CELLS

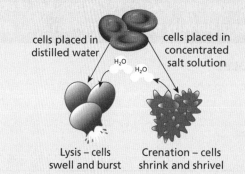

cells placed in distilled water

cells placed in concentrated salt solution

H_2O

H_2O

Lysis – cells swell and burst

Crenation – cells shrink and shrivel

Animal cells do not have a cell wall. Therefore, when too much water moves into animal cells by osmosis they will burst. This is called lysis. If too much water moves out of animal cells by osmosis they will shrink and shrivel up.

Useful substances need to move through multicellular organisms. Molecules move from areas of high concentration to areas of low concentration, along a concentration gradient. When water moves across cell membranes it is called **osmosis**. Molecules move by **diffusion** faster when: they move over a short distance with a large surface area, there is a big difference in concentration (a high **concentration gradient**), and there is a good blood supply (in animals only).

SURFACE AREA TO VOLUME RATIO

For efficient exchange cells must have a large surface area to volume ratio, this ensures cells can transport the nutrients and gases they need and release the waste the cells make. The larger a cell is the smaller its surface area to volume ratio, and so the more difficult it is for the efficient transport of nutrients, gases and wastes.

THE STRUCTURE OF PLANTS

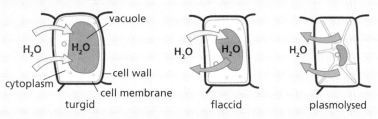

vacuole

H_2O H_2O

cytoplasm

cell wall

cell membrane

turgid

H_2O H_2O

flaccid

H_2O

plasmolysed

The cell wall and the vacuole work together to give the plant structure. When there is enough water available the vacuole pushes against the cell wall. The cell is **turgid**. When there is not enough water in the vacuole it shrinks and pulls away from the cell wall. The cell is **flaccid**. When there is not enough water in the plant it **wilts**.

Session 2   Solving the question

Plants need water for important reactions like photosynthesis.
Explain why plants wilt when they are not watered.

SOLVE

Now you have gathered the relevant information it is really important that you think carefully about how you need to use and present that information. Use the prompts below to help you.

This question asks you to **explain**. This means you need to use your scientific knowledge to write about why plants wilt without water.

Which plant sub-cellular structures are involved in the structure of the plant?

Which plant organs does the water move through from the soil to the leaf? You could list these in order.

How does the water get into the plant? Name and define the process.

How does water move from the roots to the leaves?

Plants lose water from their leaves. How does the water move out of the plant?

What happens to the structures in plant cells when they lose too much water?

Now bring together what you have written to explain why plants wilt when they do not have enough water.
You will need to include:

- what water is used for in plant cells
- how the water moves in and out of the cells
- what happens to the plant cells when the water has gone.

There are additional practice questions with the writing frames for 'Compare' and 'Describe' at the end of the book.

Session 3 Life systems

In order to move, animals need energy from respiration.
Compare how the oxygen and glucose needed for respiration enter the body and reach the cells.

THINK

This question is asking about two related journeys. You will need to use what you know about where each part of the journey happens. You should also include the cells involved and the processes that happen at each stage. You will then write how each journey is similar and different. Scribble what you remember about where oxygen and glucose come from and how and where they get into the body.

Now write down the information you will need for this specific question. Not everything you noted down before about oxygen and glucose may be relevant.

Which key idea cards can help? Deal them out next to your work and note extra information here.

Adapt *from Collins*

Don't worry about any blanks. Now have a go at the questions on Adapt© for session 3 to help you think further about this topic.

LUNGS AND GAS EXCHANGE

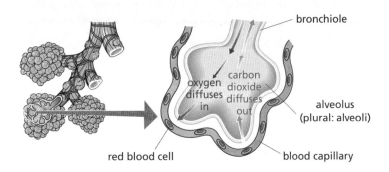

The lungs and surrounding tissues are adapted for **gas exchange**. Oxygen needed for cell respiration diffuses from the air into the blood. Carbon dioxide made by cell respiration diffuses from the blood into the air. The many small spheres of **alveoli** and the dense network of capillaries increase the surface area for diffusion. The walls of the alveoli and the capillary are one cell thick giving a short distance. The surface of the alveoli is moist allowing gases to dissolve and diffuse more easily. Blood movement and breathing maintain the concentration gradients needed for diffusion.

THE DIGESTIVE SYSTEM

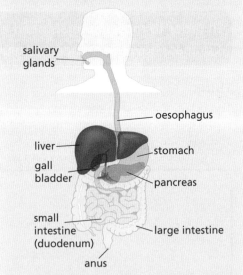

The digestive system is a group of organs that break down food into smaller soluble products which can be absorbed into the blood. Many of the organs release enzymes which act as biological catalysts, speeding up the rate of digestion.

ABSORPTION IN THE SMALL INTESTINE

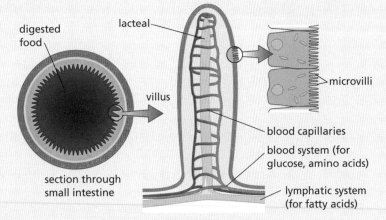

Absorption is the process by which soluble products of digestion move into the blood from the small intestine. The small intestine is about 7 m long, giving lots of time for absorption as food travels along. The inside surface (lumen) of the small intestine has **villi** which are made of cells covered in **microvilli**, which increase the surface area for absorption. Blood capillaries in the villi transport molecules away in the blood keeping the concentration gradients needed for absorption.

Life systems

ENZYME ACTION

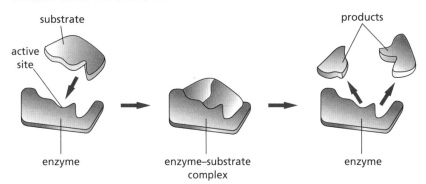

Enzymes are biological **catalysts** that speed up reactions. The **lock and key theory** is a simple model to explain how enzymes work. The **substrate** is the molecule that 'fits' into the **active site** of the enzyme like a key 'fits' into a lock. When the substrate binds to the active site an enzyme–substrate complex is formed. The enzyme then changes the substrate into the products.

Useful substances need to move through multicellular organisms. Molecules move by diffusion, osmosis or active transport. Molecules move across **specialised exchange surfaces** such as the small intestine. Exchange surfaces are adapted to allow molecules to be transported quickly. An **efficient** exchange surface has a large surface area, a thin membrane, a good blood supply (in animals only) and is well **ventilated** (for gas exchange in animals).

ENZYMES AND DIGESTION

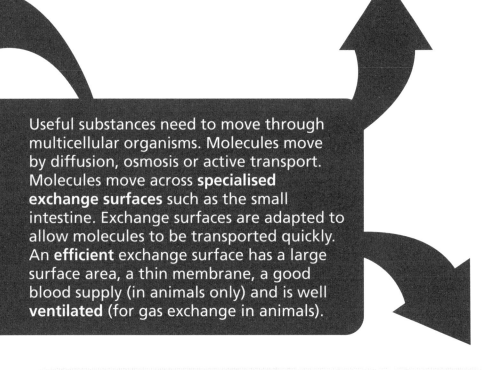

The digestive system is a connected system with the function of breaking down food for absorption. **Digestive enzymes** are released into the digestive system and work outside of cells to digest food into small soluble molecules. These small molecules are then absorbed into the blood in the small intestine. Carbohydrate digestion starts in the mouth with **amylase** (a type of carbohydrase) and then continues in the small intestine. Protein digestion starts in the stomach with **proteases** and then continues in the small intestine. Fat digestion starts in the small intestine with **lipase**.

Session 3 Solving the question

In order to move, animals need energy from respiration.
Compare how the oxygen and glucose needed for respiration enter the body and reach the cells.

SOLVE

Now you have gathered the relevant information it is really important that you think carefully about how you need to use and present that information. Use the prompts below to help you.

This question asks you to **compare**. This means you need to describe similarities and differences. You need to write about how oxygen and glucose get into and move through the body.

For oxygen: Where does oxygen come from? How does it get in? List the parts of the respiratory system that the oxygen passes through. Name the organ where the oxygen meets the blood.

Glucose is found in food that needs to be digested before it can be absorbed. Food is broken down by enzymes. What do enzymes do? What do digestive enzymes do?

The digestion of glucose happens in the digestive system. List the organs of the digestive system in order, from the mouth to the anus. Name the organ where the glucose meets the blood.

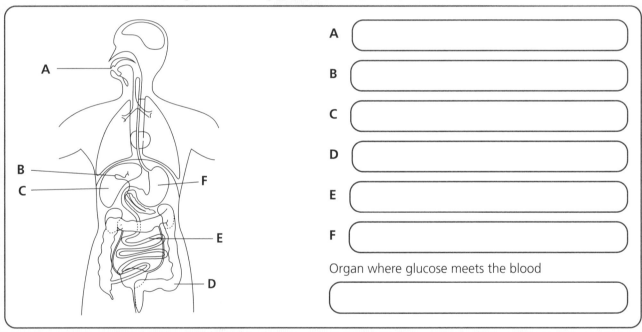

A

B

C

D

E

F

Organ where glucose meets the blood

Oxygen and glucose both diffuse into the blood. What is diffusion?

How are the organs where diffusion happens similar?

Bring together your answers in a table: How are the journeys the same? How are they different? You can use tables in the exam for comparison questions.

	Oxygen	Glucose
How substance enters the body		
How substance enters cells		

There are additional practice questions with the writing frames for 'Describe' and 'Explain' at the end of the book.

Session 4 Staying alive

Coronary heart disease is caused by lifestyle factors.
The graph shows the results of a study that looked at the effects of physical activity in men who had heart attacks.

Suggest why people with coronary heart disease get out of breath easily.

Use data from the graph to support your answer.

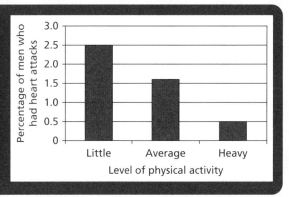

THINK

For this question you first need to think about what you know about how a healthy heart works. Next, use this in a new situation of when the heart is not working properly. Scribble what you can remember about the function (job) of the heart and what happens to the heart when you exercise.

Now write down the information you will need for this specific question. Not everything you noted down before about heart disease may be relevant.

Which key idea cards can help? Deal them out next to your work and note extra information here.

Adapt *from Collins*

Don't worry about any blanks. Now have a go at the questions on Adapt© for session 4 to help you think further about this topic.

THE RESPIRATORY SYSTEM

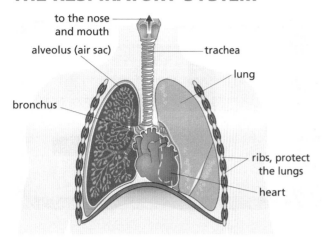

to the nose and mouth

alveolus (air sac)

trachea

lung

bronchus

ribs, protect the lungs

heart

The respiratory system is made of organs and tissues that work together to get air into and out of the body. In the lungs oxygen diffuses into the body and carbon dioxide diffuses out of the body. The **trachea** and **bronchi** are tubes through which air passes into and out of the lungs. This allows gas exchange to happen in the alveoli, which are small sacs with thin walls that are surrounded by a **capillary network**.

THE CIRCULATORY SYSTEM

cell – heart muscle cell

tissue – heart muscle

organ – the heart

organ system – the circulatory system

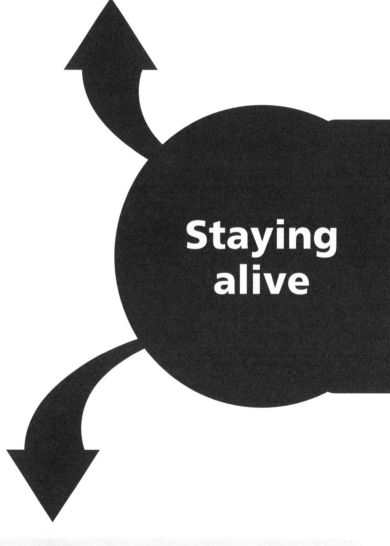

Staying alive

The **circulatory system** is a transport system made of the heart and blood vessels. It is a double system. One system pumps blood from the heart to the lungs and back. This system gets oxygen into the blood and carbon dioxide out. The other system pumps blood from the heart to the body and back. This provides cells with the oxygen and nutrients they need (for respiration) as well as taking away the carbon dioxide (produced in respiration).

STRUCTURE OF THE HEART

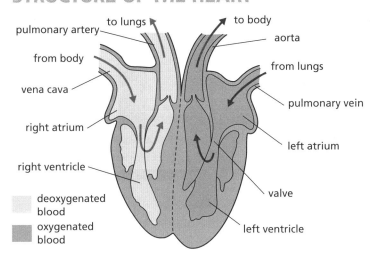

- pulmonary artery
- to lungs
- to body
- aorta
- from body
- from lungs
- vena cava
- pulmonary vein
- right atrium
- left atrium
- right ventricle
- valve
- deoxygenated blood
- oxygenated blood
- left ventricle

The heart is an organ made of specialised **cardiac muscle** tissue. The right side of the heart receives blood from the body through the **vena cava** into the right **atrium**. Then, the right **ventricle** pumps blood to the lungs through the **pulmonary artery**. In the lungs, gas exchange happens. The oxygenated blood then returns to the left atrium through the **pulmonary vein**. The left ventricle then pumps the blood to the body through the **aorta**, this has the thickest wall because it carries blood under very high pressure to the rest of the body. The heart rate can increase and decrease depending on what the body needs. When we exercise more oxygen is needed so the heart pumps the blood faster.

Cells are the building blocks of living organisms. Cells with similar structure and function are grouped together into **tissues**. Tissues with a similar function work together to form **organs**. Organs are grouped into organ systems which have an overall function. For example, they move substances around the body. Organ systems work together to keep living organisms alive. Damage to cells, tissues or organs within body organ systems can cause disease or death.

BLOOD AND BLOOD VESSELS

Parts of blood

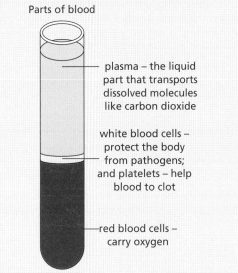

- plasma – the liquid part that transports dissolved molecules like carbon dioxide
- white blood cells – protect the body from pathogens; and platelets – help blood to clot
- red blood cells – carry oxygen

The circulatory system carries blood around the body. Blood is carried in three main types of vessel – arteries, veins and capillaries. These vessels are specialised to their function, as shown in the table. **Coronary arteries** supply the heart muscle with blood. **Coronary heart disease** is caused by damage to the coronary arteries or by fatty deposits forming in the vessels.

Arteries	Veins	Capillaries
thick, elastic wall, not **permeable**; small lumen	thin wall, not permeable; large lumen; valve – prevents backflow of blood	wall is one cell thick and permeable to allow oxygen to diffuse into tissues
carry blood away from the heart	carry blood to the heart	carry blood from arteries to tissues and then join up with veins
carry blood under high pressure with a pulse	carry blood under low pressure, blood flows smoothly	blood pressure falls and pulse disappears as the blood travels through the capillaries

Session 4 Solving the question

Coronary heart disease is caused by lifestyle factors. The graph shows the results of a study that looked at the effects of physical activity in men who had heart attacks.

Suggest why people with coronary heart disease get out of breath easily.

Use data from the graph to support your answer.

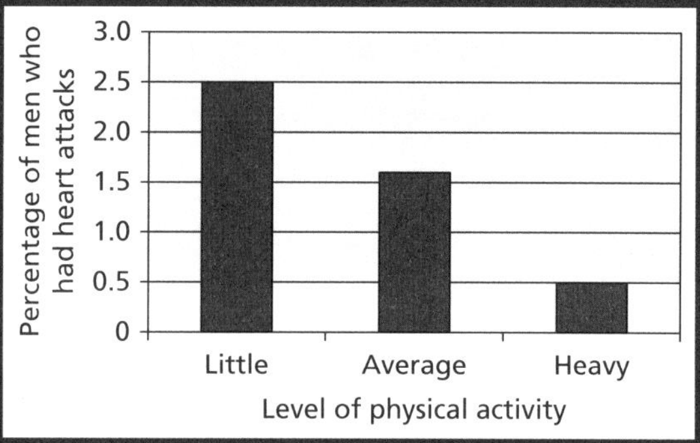

SOLVE

Now you have gathered the relevant information it is really important that you think carefully about how you need to use and present that information. Use the prompts below to help you.

This question asks you to **suggest**. This means you need to apply what you know to a new situation.

The heart is part of the circulatory system. What are the three types of blood vessel found in the circulatory system?

When you exercise your muscle cells need more energy from respiration. What is needed for respiration?

Where do the substances needed for respiration come from?

How does the circulatory system move the substances to the muscles?

Coronary heart disease is caused by lifestyle factors. Which lifestyle factors can lead to coronary heart disease?

What does the data in the chart show you? Describe the results. Remember to include numbers from the graph.

What happens to your breathing and heart rate when you exercise?

Why does this happen?

Now bring together what you have written about how the heart works and what happens when you exercise to answer the question:

Suggest why people with coronary heart disease get out of breath more easily.

There are additional practice questions with the writing frames for 'Compare' and 'Evaluate' at the end of the book.

Session 5 Nervous control

Humans have reflexes to protect them from harm.
Show how a reflex arc makes you quickly move your hand away from something hot.

THINK

This question asks you to **show** how the reflex arc works. You need to write the steps of the reflex arc working to move a hand away from something hot. Scribble or draw what you can remember about the nervous system, reflex actions and the reflex arc.

Now write down the information you will need for this specific question. Not everything you noted down before about reflexes may be relevant.

Which key idea cards can help? Deal them out next to your work and note extra information here.

Adapt *from Collins*

Don't worry about any blanks. Now have a go at the questions on Adapt© for session 5 to help you think further about this topic.

THE NERVOUS SYSTEM

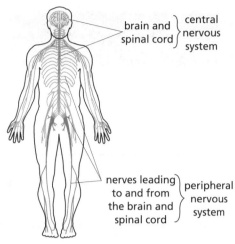

brain and spinal cord } central nervous system

nerves leading to and from the brain and spinal cord } peripheral nervous system

The nervous system is made of the **central nervous system** (CNS) and the **peripheral** nervous system. The CNS is made of the brain and the spinal cord. The CNS coordinates the response to a **stimulus** – a change in the environment.

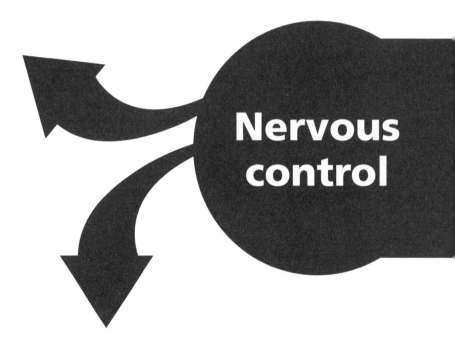

Nervous control

STRUCTURE OF NERVE CELLS

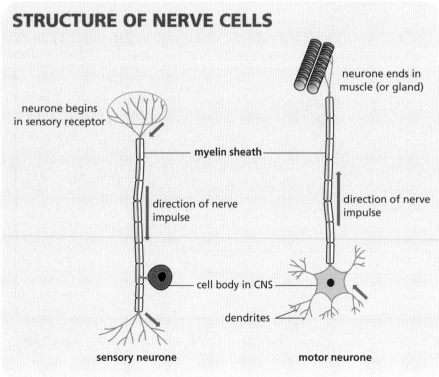

neurone begins in sensory receptor

neurone ends in muscle (or gland)

myelin sheath

direction of nerve impulse

direction of nerve impulse

cell body in CNS

dendrites

sensory neurone

motor neurone

Nerve cells (also called **neurones**) are the specialised cells of the nervous system. Nerve cells can transmit electrical impulses. At the ends of the nerve cell are fine branches called dendrites that help neurones connect to other cells. Nerve cells are also specialised to be very long and have a fatty layer of myelin. The length helps the nerve cells transmit impulses around the body, and the myelin insulates the nerve so it can transmit information quickly. **Sensory neurones** carry information about changes in the environment. **Motor neurones** carry information about how to respond to the environment.

COORDINATING RESPONSES

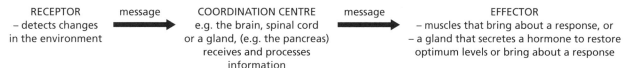

RECEPTOR
– detects changes
in the environment

message →

COORDINATION CENTRE
e.g. the brain, spinal cord
or a gland, (e.g. the pancreas)
receives and processes
information

message →

EFFECTOR
– muscles that bring about a response, or
– a gland that secretes a hormone to restore
optimum levels or bring about a response

Sensory neurones connect receptors to the CNS. Receptors are special areas found in our sense organs which detect a stimulus. For example, light receptors are found in our eyes; temperature receptors are found in our skin. Motor neurones connect the CNS to effectors. Effectors are muscles or glands and allow our bodies to respond to changes in the environment. For example, when we smell food the salivary glands in our mouth release saliva.

Our bodies have control systems that communicate with parts of the body to make changes. These changes keep our internal environment stable so that the processes that keep us alive work properly. In nervous control, changes are detected by **receptors**. Receptors send impulses to the **coordination centre** (brain or spinal cord) via an electrical signal in nerves. The brain then coordinates an action. The action is sent in nerves to an **effector** organ. Nervous control is fast- and short-acting.

THE REFLEX ARC

- 4 relay neurone
- spinal cord
- 5 motor neurone
- 6 effector (biceps muscle)
- 3 sensory neurone
- 2 receptors in skin
- 7 response – hand moved away
- 1 hot plate (stimulus)

Reflex actions are rapid, automatic responses to a stimulus. In simpler organisms they are the basis of behaviour. In humans they stop us from getting hurt. Reflex actions include closing our eyes if an object comes near to them, and the grasping reflex of a baby gripping a finger. The diagram shows a reflex action to touching a hot plate. The stimulus (touching the plate) is detected by a receptor (in the skin). The impulse is then sent to the CNS through a sensory neurone. The **relay neurones** in the CNS bring about a response and the impulse is sent to an effector (the biceps muscle) through a motor neurone. The reflex action makes sure the person quickly moves their hand away from the hot plate. The stimulus relay neurones in the CNS (*normally the spinal cord*).

Session 5 Solving the question

Humans have reflexes to protect them from harm.
Show how a reflex arc makes you quickly move your hand away from something hot.

SOLVE

Now you have gathered the relevant information it is really important that you think carefully about how you need to use and present that information. Use the prompts below to help you.

This question asks you to **show** how the reflex arc works. This means you need to write a series of steps that move the hand away from something hot. You could also draw a diagram to answer a 'show' question.

Which organs are in the central nervous system?

Neurones or nerve cells are specialised to conduct nerve impulses. How are neurones specialised?

What are the three types of neurone found in a reflex arc?

Draw the reflex arc and list the steps. Make sure you include the names of each type of neurone.

Now bring together all the steps in order and link the fast reflex to the need to move quickly away from the heat. When the hand touches something hot it needs to move quickly because …

The hand moves quickly because this is a reflex action. The steps of a reflex action are...

What is another example of a reflex action?

There are additional practice questions with the writing frames for 'Compare' and 'Show' at the end of the book.

Session 6 Hormonal control

The graph shows how the amount of glucose in a healthy person's bloodstream changes after eating a meal.

a. Explain how your body controls blood sugar, by homeostasis, using hormones.

b. Diabetes is a condition in which blood sugar is uncontrolled.

Explain how the shape of the graph would look different in someone with diabetes.

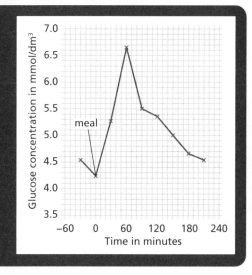

THINK

This question asks you to **explain** how homeostasis works. This means you will need to write a detailed answer that includes all the stages of the control of blood sugar. You should include all the organs involved. Scribble or draw what you can remember about blood sugar control and homeostasis.

Now write down the information you will need for this specific question. Not everything you noted down before about control of blood sugar may be relevant.

Which key idea cards can help? Deal them out next to your work and note extra information here.

Don't worry about any blanks. Now have a go at the questions on Adapt© for session 6 to help you think further about this topic.

THE ENDOCRINE SYSTEM

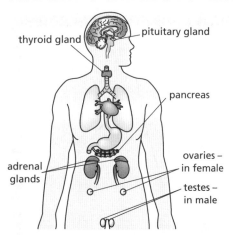

The **endocrine system** is made up of glands that make specific chemical messengers called hormones which are released into the blood.

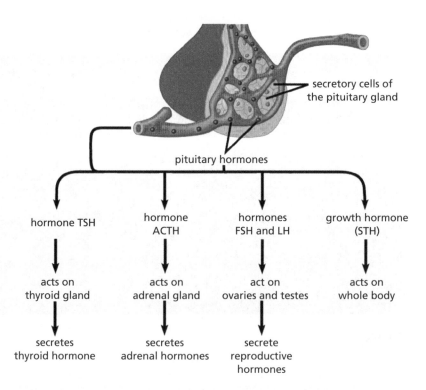

pituitary hormones			
hormone TSH	hormone ACTH	hormones FSH and LH	growth hormone (STH)
acts on thyroid gland	acts on adrenal gland	act on ovaries and testes	acts on whole body
secretes thyroid hormone	secretes adrenal hormones	secrete reproductive hormones	

The pituitary gland sits just below the brain and is described as the 'master gland'. The pituitary secretes some hormones that have a direct effect on target organs; for example, growth hormone. The pituitary also secretes hormones that have an indirect effect by acting on other glands to secrete their hormones into the blood.

HOMEOSTASIS

Homeostasis is the control of conditions inside the body between narrow limits. This control is automatic and is carried out by the nervous and endocrine systems. Water levels and body temperature have to be kept stable for the optimum conditions for enzymes to work.

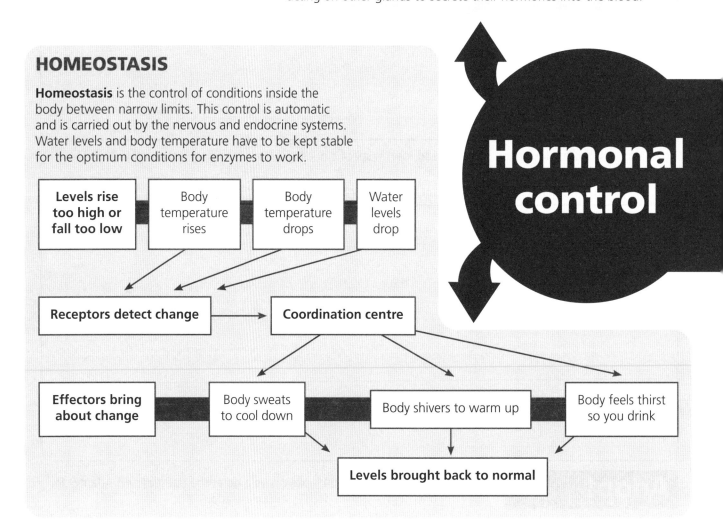

Hormonal control

| **Levels rise too high or fall too low** | Body temperature rises | Body temperature drops | Water levels drop |

Receptors detect change → **Coordination centre**

| **Effectors bring about change** | Body sweats to cool down | Body shivers to warm up | Body feels thirst so you drink |

Levels brought back to normal

CONTROLLING BLOOD SUGAR

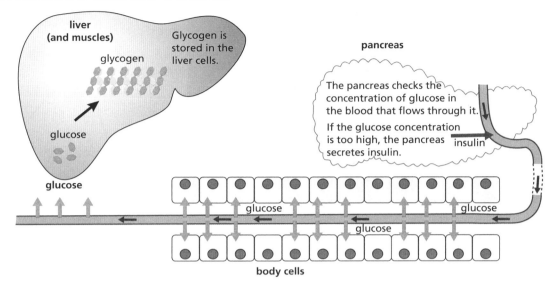

Blood sugar (glucose) is controlled by the **pancreas**. The pancreas detects changes and releases the hormone insulin if sugar levels in the blood get too high. Insulin makes glucose go into cells to be used in cell respiration, or into the liver or muscles to be stored as the carbohydrate **glycogen**. People with diabetes cannot control their blood glucose levels. In **Type 1 diabetes** the pancreas cannot produce enough insulin. This causes high blood sugar levels that are treated by injecting insulin. In **Type 2 diabetes** the cells no longer respond to insulin and sometimes the body cannot make enough. Type 2 diabetes can be treated by a carbohydrate-controlled-diet, increased exercise and, if necessary, medication, including insulin. Obesity is a risk factor for Type 2 diabetes.

Our bodies have control systems that communicate with parts of the body to make changes. These changes keep our internal environment stable so that the processes that keep us alive work properly. In hormonal control, receptors sense when a **hormone** needs to be produced. Hormones are secreted by glands and travel through the blood to the organs where they have an effect. Hormones act more slowly than nerves, but their effects last longer.

PUBERTY IN FEMALES

Last day
27th 28th
26th
25th   1st day 2nd
24th   3rd
menstruation
23rd   egg dies   4th
if not fertilised
22nd   uterus   5th
lining
is shed
21st   6th

20th   7th
uterus lining
starts to repair and 8th
ovulation   grow again
19th uterus lining
continues to thicken   9th
18th
egg is released   10th
17th   from ovary 11th
16th   12th
15th 14th 13th

During puberty hormones are released which help the body develop **secondary sex characteristics**. In females the **menstrual cycle** begins. **Follicle stimulating hormone** (FSH) causes an egg to mature in the ovary. **Luteinising hormone** (LH) causes the ovary to release the mature egg. **Oestrogen** and **progesterone** both help to repair and maintain the lining of the **uterus**. To prevent pregnancy **contraception** can be used. Barrier methods (condom and diaphragm) prevent the sperm from reaching the egg. Hormonal methods (implant, patch, pill) stop eggs from being released by inhibiting FSH.

Session 6 Solving the question

The graph shows how the amount of glucose in a healthy person's bloodstream changes after eating a meal.

a. Explain how your body controls blood sugar, by homeostasis, using hormones.

b. Diabetes is a condition in which blood sugar is uncontrolled.

   Explain how the shape of the graph would look different in someone with diabetes.

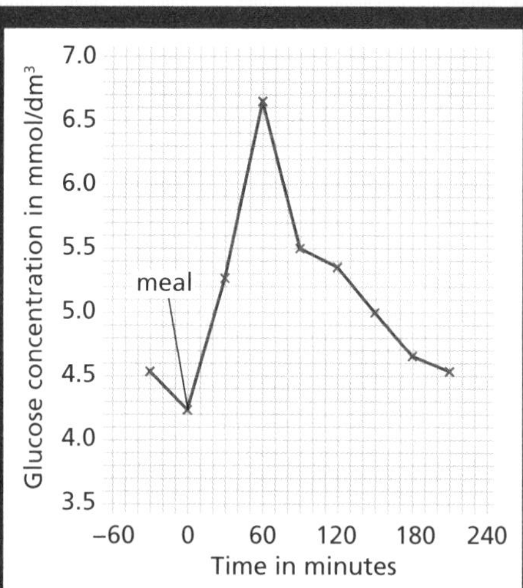

SOLVE

Now you have gathered the relevant information it is really important that you think carefully about how you need to use and present that information. Use the prompts below to help you.

This question asks you to **explain**. This means you need to use what you know about how hormones work to write how blood sugar is controlled.

For part (a): Where does the sugar in the blood come from?

For part (a): How is glucose (sugar) removed from the blood?

For part (a): How does the body know if there is too much glucose in the blood?

For part (a): Now put it all together to explain how your body controls blood sugar.

For part (b): Look at the graph. Describe the shape of the graph in your own words.

For part (b): People who have a poor diet may develop Type 2 diabetes. What happens to blood sugar levels in someone with diabetes?

For part (b): Draw a sketch on the graph in the question to show what would happen to blood sugar in someone with diabetes. Remember that your sketch is a good guess or estimate of how the line would look. How is your sketch different from the graph for the healthy person?

There are additional practice questions with the writing frames for 'Compare' and 'Define' at the end of the book.

Session 7 Disease and immunity

> Your immune system stops infections caused by pathogens (harmful microbes).
> Explain how vaccines trigger this process and how this protects you.

THINK

This question asks you to **explain** how vaccines work. This means you will need to write a detailed answer that includes all the steps in the process of developing immunity. Scribble or draw what you can remember about infectious diseases, vaccines and how the body reacts to infection.

Now write down the information you will need for this specific question. Not everything you noted down before about the immune system may be relevant.

Which key idea cards can help? Deal them out next to your work and note extra information here.

Adapt *from Collins*

Don't worry about any blanks. Now have a go at the questions on Adapt© for session 7 to help you think further about this topic.

THE IMMUNE SYSTEM

The **immune system** has two parts.
1. Barriers, which stop pathogens entering the body. 2. White blood cells which attack pathogens if they get into the body.

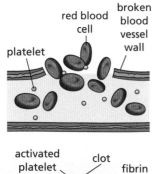

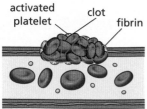

Barriers like the skin, hair and mucus in the nose, and stomach acid stop pathogens entering the body. If your skin is damaged, the blood quickly clots to reform the skin barrier.

1 A phagocyte moves towards a bacterium.

2 The phagocyte pushes a sleeve of cytoplasm outwards to surround the bacterium.

3 The bacterium is now enclosed in a vacuole inside the cell. It is then killed and digested by enzymes.

Phagocytes are a type of white blood cell which ingest many different types of pathogen.

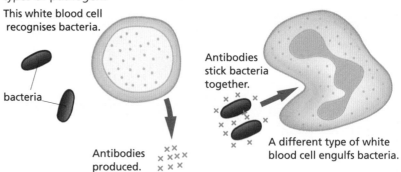

This white blood cell recognises bacteria.

bacteria

Antibodies produced.

Antibodies stick bacteria together.

A different type of white blood cell engulfs bacteria.

Other white blood cells called **lymphocytes** are **specific** to a particular pathogen. Lymphocytes make **antibodies** which stick bacteria together and **antitoxins** which neutralise toxins (poisons) made by pathogens.

Disease and immunity

PREVENTING DISEASE

When live pathogens enter the body, lymphocytes instantly recognise them and respond more quickly to the infection.

Lymphocytes produce antibodies against the inactive or dead pathogens.

Lymphocytes remember the shape of the antigen.

Antibody response

Time

Vaccination

Infection by pathogen

Antibiotics are medicines like **penicillin** which can be given to patients with a bacterial infection. Antibiotics cannot kill **viruses**. Some bacteria are building resistance which means the antibiotics can no longer kill them. **Vaccines** can be used to prevent infection by giving someone small quantities of dead or inactive forms of a pathogen. This stimulates an **immune response**. An immune response means the body quickly makes antibodies if an infection occurs. Vaccines can be used for many different types of pathogen, including viruses.

PATHOGENS

Pathogens are organisms or viruses that cause a disease. Pathogens can be **eukaryotes** like fungi and protists, or **prokaryotes** like bacteria, or viruses.

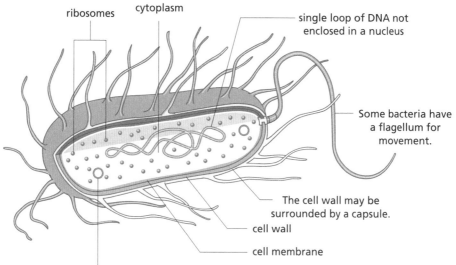

ribosomes   cytoplasm

single loop of DNA not enclosed in a nucleus

Some bacteria have a flagellum for movement.

The cell wall may be surrounded by a capsule.

cell wall

cell membrane

Small ring of DNA called a plasmid (one or more in a cell). Genes in the plasmids can give the bacterium advantages such as antibiotic resistance.

Bacteria are prokaryotic cells which are much smaller than human cells and can be pathogenic. They have different proteins and our body recognises any cell that is not a human cell and makes antibodies to attach to them.

Viruses are very small pathogens which replicate (make copies) inside host cells to make new copies of themselves to be released. This causes damage to the body cells.

Pathogens are microbes that cause disease by destroying body cells and/or producing toxins. Pathogens have different proteins from body cells. **White blood cells** recognise these proteins as **foreign**. The body makes white blood cells that can attach to the foreign proteins and destroy them. Some of the matching white blood cells remain in the body after the pathogen is destroyed. If the pathogen enters the body again, the white blood cells can multiply quickly and prevent disease.

PATHOGENS AND THEIR DISEASES

Pathogen	Type	Symptoms	How it spreads
Measles	virus	Fever and red skin rash, can be fatal	Inhalation of droplets
HIV	virus	Flu-like illness and eventually an ineffective immune system	Through body fluids like blood and semen
Tobacco mosaic virus	virus	Mosaic pattern of discolouration on plants, decreasing photosynthesis	Direct contact
Salmonella	bacterium	Vomiting, diarrhoea and abdominal cramps	Eating food infected with the bacterium
Gonorrhoea	bacterium	Thick yellow or green discharge from the vagina or penis	Sexual contact
Rose black spot	fungus	Purple or black spots develop on leaves, decreasing photosynthesis	Carried by wind or water
Malaria	protist	Fever, can be fatal	Carried by mosquitoes

Session 7 Solving the question

Your immune system stops infections caused by pathogens (harmful microbes).
Explain how vaccines trigger this process and how this protects you.

SOLVE

Now you have gathered the relevant information it is really important that you think carefully about how you need to use and present that information. Use the prompts below to help you.

This question asks you to use what you know about how the immune system works. You need to **explain** how vaccines protect you from infectious diseases.

What is the general term for an organism or virus that causes infectious diseases? What kind of organisms other than viruses can cause infectious diseases?

What are the barriers that stop pathogens getting into the body?

If pathogens do get in, how does the body recognise them?

Which cells are involved in fighting the pathogens?

What do these cells do?

Some of the white blood cells that have fought off an infection remain in the body. How does this help you if you are infected with the same pathogen again?

Why does this only work for that specific pathogen?

Vaccines contain a tiny part of a pathogen, or a pathogen that has been damaged so it does not cause disease. How does a vaccine stop you getting ill?

There are additional practice questions with the writing frames for 'Define' and 'Explain' at the end of the book.

Session 8 Life cycles

The materials that living things are made of have existed on Earth for millions of years.

Describe how carbon moves through living things via respiration, photosynthesis and decomposition.

Include the conditions needed for each process.

THINK

This question asks you to **describe** how carbon is used and passed on by living things. You will need to think about each process separately and as a cycle. Scribble or draw what you can remember about respiration, photosynthesis, food chains and decomposition (rotting).

Now write down the information you will need for this specific question. Not everything you noted down before about the carbon cycle may be relevant.

Which key idea cards can help? Deal them out next to your work and note extra information here.

Adapt *from Collins*

Don't worry about any blanks. Now have a go at the questions on Adapt© for session 8 to help you think further about this topic.

PHOTOSYNTHESIS

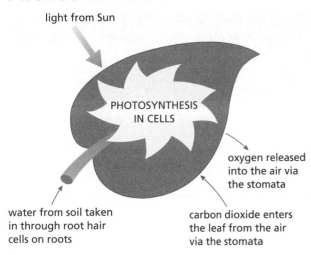

light from Sun

PHOTOSYNTHESIS
IN CELLS

oxygen released
into the air via
the stomata

water from soil taken
in through root hair
cells on roots

carbon dioxide enters
the leaf from the air
via the stomata

used to make cellulose
for plant cell walls

used in cell respiration

stored as starch

used to make
proteins for
plant growth
and repair

stored as
fats and oils

glucose

photosynthesis

oxygen

used in respiration

released into air

The sugars produced are stored or used by the plant to grow. Plants also use the sugar in cell respiration along with the oxygen produced in photosynthesis.

Photosynthesis is one of the most important chemical reactions in living things as it is the basis for most **food chains**. Plants use the energy from sunlight to turn water (H_2O) and carbon dioxide (CO_2) into simple sugars such as glucose ($C_6H_{12}O_6$).

water + carbon dioxide $\xrightarrow{\text{light and chlorophyll}}$ glucose + oxygen (O)

RESPIRATION

Respiration happens in cells. Animals use the sugar from their food, and oxygen for **aerobic respiration**. The equation is:

glucose + oxygen → carbon dioxide + water

Without oxygen, **anaerobic respiration** takes place in muscles:

glucose → lactic acid

This does not give as much energy as aerobic respiration.

In plants and yeast, the equation for anaerobic respiration is:

glucose → ethanol + carbon dioxide

Anaerobic respiration in yeast cells is called **fermentation**. Fermentation is used in the making of bread and alcoholic drinks.

Energy transferred by respiration supplies all the energy needed for living processes.

Life cycles

ECOSYSTEMS

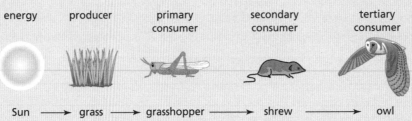

| energy | producer | primary consumer | secondary consumer | tertiary consumer |

Sun ⟶ grass ⟶ grasshopper ⟶ shrew ⟶ owl

Ecosystems are a web of interactions between living organisms (**biotic factors**) and the **abiotic** (non-living) parts of their surroundings. Within an ecosystem living things **compete** for resources like space and food, but also depend on other species for food and shelter. This is **interdependence**. Some organisms need other species to help them reproduce. For example, flowers need bees for **pollination**. **Biodiversity** is the variety of all the **species** in an ecosystem. Changes to part of an ecosystem can affect all other parts.

NUTRIENT CYCLING

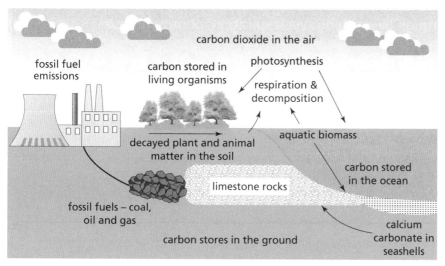

Nutrients cycle through the biotic and abiotic parts of an ecosystem through feeding and decomposition. Nutrients in the bodies of organisms are recycled by bacteria and other decomposers in the soil. The nutrients released are available for other organisms to use. Carbon found in living organisms is released to the atmosphere by cell respiration. Carbon is then returned to the food chain through photosynthesis in plants.

The water cycle provides fresh water for animals and plants on land where it eventually drains into the sea. Water evaporates back into clouds and is then returned to the land through **precipitation** (rainfall).

All the substances that living things are made of are recycled. Plants fix carbon dioxide from the air and make it into sugar using light in photosynthesis. The sugar is used to build plant bodies. Animals use the carbon fixed by plants to build their own bodies. Animals release energy from sugar from plants using **respiration**. When organisms die, the substances in their bodies are **decomposed** by **microorganisms** so they can be reused.

BIODIVERSITY

Biodiversity is essential to ensuring the stability of ecosystems by giving more food options to each each species. The future of the human species relies on us maintaining high levels of biodiversity.

HUMAN INTERACTIONS

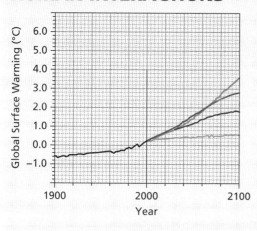

Humans have both positive and negative interactions with the environment. Human land use, such as building, quarrying, farming and waste dumping, has led to **deforestation** and habitat destruction. The increased release of carbon dioxide and methane into the environment has led to **global warming**. This has resulted in a climate crisis which may result in increased temperatures, sea level rise and loss of biodiversity. Scientists make predictions about how much the Earth will warm (the graph shows four different models of global warming) and these are evaluated using **peer review**. Population growth has led to an increased need for resources which has increased the amount of pollution released. However, humans are now taking action to reduce these negative impacts on biodiversity. Scientists and concerned citizens have started breeding programmes for **endangered species**, and encouraged protection and regeneration of rare habitats.

Session 8   Solving the question

The materials that living things are made of have existed on Earth for millions of years.
Describe how carbon moves through living things via respiration, photosynthesis and decomposition.
Include the conditions needed for each process.

SOLVE

Now you have gathered the relevant information it is really important that you think carefully about how you need to use and present that information. Use the prompts below to help you.

This question asks you to **describe** how carbon is used and passed on by living things. You will need to remember the processes of respiration, photosynthesis and decomposition and write how they are all linked.

Plants are the first user of carbon, so we will start there. How do plants use carbon? What do they make it into? Write an equation for the process.

Carbon can be passed on by the plant dying or being eaten. What happens to a plant if it dies? What are the conditions for decomposition?

Carbon is passed along the food chain. Draw and label a food chain, naming each level. Use the labels: producer, primary consumer, secondary consumer, tertiary consumer.

Glucose contains carbon. Carbon in glucose is used in the process of respiration to release energy in plants and animals. Oxygen is needed, and carbon dioxide and water are produced. Show this as a word equation.

Carbon can be released from fossil fuels when they are burned.
Now sum up how carbon gets into organisms, and how it leaves. You could draw a labelled diagram.

There are additional practice questions with the writing frames for 'Describe' at the end of the book.

Session 9 Reproduction and inheritance

The organisms alive today are very different from the organisms that lived millions of years ago. Peppered moths are either white and speckled, or black. During the daytime peppered moths rest on branches of trees. In the 19th century the bark of these trees became covered in black soot.

Show how peppered moths got darker in colour through natural selection when the trees they lived on became sooty.

THINK

This question needs you to know how organisms can be different from each other and how they can change over time. Scribble down anything you can remember about how and why organisms are different and how characteristics are passed on between generations.

Now write down the information you will need for this specific question. Not everything you noted down before about natural selection may be relevant.

Which key idea cards can help? Deal them out next to your work and note extra information here.

Adapt *from Collins*

Don't worry about any blanks. Now have a go at the questions on Adapt© for session 9 to help you think further about this topic.

MEIOSIS

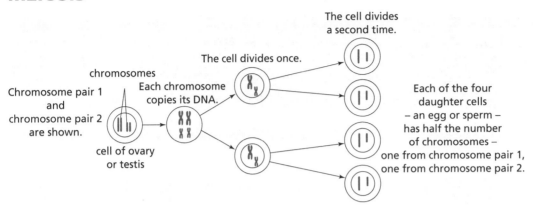

The cell divides a second time.

The cell divides once.

chromosomes

Chromosome pair 1 and chromosome pair 2 are shown.

Each chromosome copies its DNA.

cell of ovary or testis

Each of the four daughter cells – an egg or sperm – has half the number of chromosomes – one from chromosome pair 1, one from chromosome pair 2.

Meiosis is a type of cell division in reproductive organs that makes four **gametes** (sex cells). Gametes are eggs and sperm in humans, and pollen and egg cells in flowering plants. The original cell with chromosomes in pairs divides twice to form four gametes. Each gamete has a single set of chromosomes and is genetically different (non-identical) to the other gametes produced. Meiosis is a way of introducing **variation** into a species.

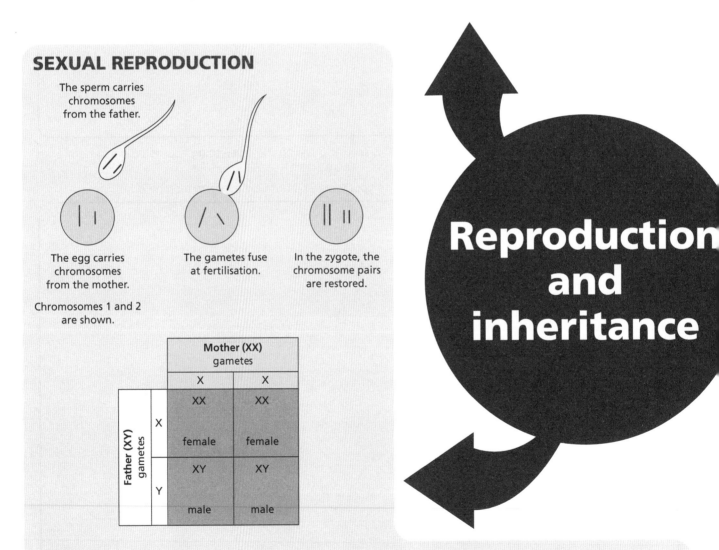

SEXUAL REPRODUCTION

The sperm carries chromosomes from the father.

The egg carries chromosomes from the mother.

Chromosomes 1 and 2 are shown.

The gametes fuse at fertilisation.

In the zygote, the chromosome pairs are restored.

		Mother (XX) gametes	
		X	X
Father (XY) gametes	X	XX female	XX female
	Y	XY male	XY male

Reproduction and inheritance

At **fertilisation** the gametes (sex cells) from each parent fuse to make a fertilised egg. The sex of the baby is determined by the **sex chromosomes**. Egg cells will always have an X chromosome, and sperm cells can have either an X or a Y chromosome. If a sperm with an X chromosome fuses with the egg the baby born will be female with XX sex chromosomes. If the sperm has a Y chromosome the baby born will be male with XY sex chromosomes.

Single-celled organisms can reproduce by **asexual reproduction** but this does not produce variation.

EVOLUTION BY NATURAL SELECTION

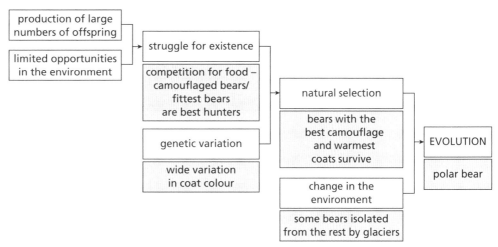

Evolution is the change in the inherited characteristics of a population over time through **natural selection**. When there is a change in the environment, variation in a species means some individuals are better **adapted** to the new conditions. These individuals therefore survive and reproduce. They pass on the favourable characteristics. Over time the species changes. If two groups of a species are isolated in two different environments natural selection will lead to different adaptations being selected. Eventually, the species will differ so much they become different species and can no longer **interbreed**. The flow chart shows how polar bears may have evolved from brown bears as they were separated by ice.

Fossils provide evidence for evolution by showing changes in organisms that lived before in different aged rocks. Artificial selection (picking individuals with specific traits to breed together) has shown how species can change.

Characteristics are caused by the code in **DNA**. In eukaryotes DNA is stored in the nucleus of cells. When plants and animals reproduce they make gametes (sex cells) that have half the DNA of body cells. The DNA is different in each gamete and is chosen randomly. DNA can change by **mutations** when cells divide. These processes mean that individuals have lots of small differences from each other. Some of the differences help organisms survive.

VARIATION

Darker moth   Light moth

Variation is important to ensure a species survives changes in the environment. Some individuals are better suited to the new environment. Genetic variation comes from mutation, meiosis and sexual reproduction. Many mutations in the DNA may have no effect. Some mutations could result in a new **phenotype** (characteristic) being produced, such as peppered moths having darker wings.

Session 9 Solving the question

The organisms alive today are very different from the organisms that lived millions of years ago. Peppered moths are either white and speckled, or black. During the daytime peppered moths rest on branches of trees. In the 20th century the bark of these trees became covered in black soot.

Show how peppered moths got darker in colour through natural selection when the trees they lived on became sooty.

SOLVE

Now you have gathered the relevant information it is really important that you think carefully about how you need to use and present that information. Use the prompts below to help you.

This question asks you to **show** how organisms change over time. This means you need to write down the steps of natural selection.

In the original population of moths, most moths were light coloured, and a few were dark. Describe why some organisms look different.

When the trees got sooty, more of the dark-coloured moths survived. Why?

Camouflage like this is an adaptation to not being eaten. Animals and plants have many adaptations that help them survive. Write or draw four examples below.

More of the next generation of moths were darker coloured. Why?

Over time the whole population became darker coloured. Bring together the steps that led to this.

There are additional practice questions with the writing frames for 'Show' and 'Suggest' at the end of the book.

Session 10 Genetics and genetic engineering

Polydactyly is an inherited condition caused by a dominant allele.

A pregnant woman does not have polydactyly, but her husband Albert does.

Albert's mother does not have polydactyly.

Determine the possible genotypes and phenotypes for their child.

Complete the Punnett square shown to work out your answer.

Use D for the dominant allele and d for the recessive allele.

	Mother (dd) gametes	
	d	d
Father (Dd) gametes D		
d		

THINK

This question asks you to work out the probability of a child having an inherited condition. Without looking at the knowledge organiser, what do you remember about the inheritance of genetic conditions and the words we use to talk about genetics? Scribble down what you know about genetic material, how we write alleles and genetic diagrams.

Now write down the information you will need for this specific question. Not everything you noted down before about genetics may be relevant.

Which key idea cards can help? Deal them out next to your work and note extra information here.

Adapt *from Collins*

Don't worry about any blanks. Now have a go at the questions on Adapt© for session 10 to help you think further about this topic.

MITOSIS AND STEM CELLS

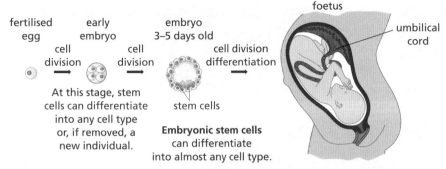

fertilised egg

early embryo

cell division

cell division

embryo 3–5 days old

cell division differentiation

foetus

umbilical cord

At this stage, stem cells can differentiate into any cell type or, if removed, a new individual.

stem cells

Embryonic stem cells can differentiate into almost any cell type.

Cells divide during the stage of the **cell cycle** called mitosis. First the genetic material doubles and then it separates into two identical cells. Mitosis provides cells for growth and repair. **Undifferentiated** cells that can develop into a number of cells are called **stem cells**. **Embryonic stem cells** can produce any type of cell. **Adult stem cells** are more limited and can only produce a small number of different cell types. In **therapeutic cloning** stem cells with the same genes as the patient are used to replace damaged tissues.

Genetics and genetic engineering

DNA

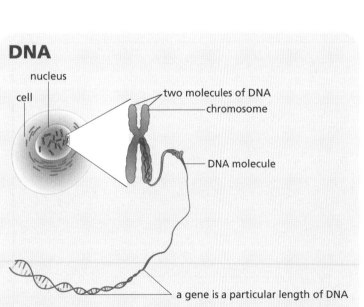

nucleus

cell

two molecules of DNA

chromosome

DNA molecule

a gene is a particular length of DNA

The **genome** (all the genetic material) of a human cell is made of DNA, which is in the shape of a **double helix**. The DNA coils up to form **chromosomes**. A small section of DNA is called a **gene**. Each gene carries the code for a specific protein.

GENES

the genes controlling a certain characteristic are at the same location on each chromosome of the pair

chromosome pair 1   chromosome pair 2   chromosome pair 3   → 23 pairs

Humans have 23 pairs of chromosomes which carry versions of genes called **alleles**. If someone has two copies of the same allele they are **homozygous** for that gene. If they have two different alleles, they are **heterozygous** for that gene. The alleles someone has is their **genotype** and the outward characteristics of their genotype is their phenotype.

GENETICS

		Mother (Ff) gametes	
		F	f
Father (Ff) gametes	F	FF unaffected	Ff unaffected (carrier)
	f	Ff unaffected (carrier)	ff cystic fibrosis

Dominant alleles are always expressed even if someone only has one copy, like **polydactyly** (a trait where people have extra fingers or toes). **Recessive** alleles can only be expressed if someone has two copies, as in **cystic fibrosis**. You can use a **Punnett square** diagram to predict the **probability** (chance) of the phenotypes of children. You will need to know the genotype of the parents. In the example, both parents are heterozygous (Ff) so they have a 25% chance of having a child with cystic fibrosis (ff).

Characteristics are caused by the code in DNA; the form and function of bodies or organisms are controlled by the code of the genetic material, such as DNA. In eukaryotes DNA is stored in the **nucleus** of cells. When plants and animals reproduce they make gametes (sex cells) that have half the DNA of body cells. The DNA is different in each gamete and is chosen randomly. Exact copies of cells are made by division when plants and animals grow. This is called **mitosis**. Humans use **genetic engineering** to make products, plants and animals for food and medicine.

GENETIC ENGINEERING

For thousands of years people have bred animals or plants which had useful traits. This is called **selective breeding**. After many generations of selective breeding the useful trait is found in all of the offspring. Genetic engineering is a much more direct way and allows the addition of useful genes from one species into another. Genetically modified or **GM crops** are formed by genetic engineering. This has allowed farmers to grow plants that give bigger **yields** and are resistant to disease and attack from insects. Some people are worried about potential harmful effects to human health and the environment. Genetic engineering might also be used to treat some inherited disorders like cystic fibrosis.

Session 10   Solving the question

> Polydactyly is an inherited condition caused by a dominant allele.
>
> A pregnant woman does not have polydactyly, but her husband Albert does.
>
> Albert's mother does not have polydactyly.
>
> Determine the possible genotypes and phenotypes for their child.
>
> Complete the Punnett square shown to work out your answer.
>
> Use D for the dominant allele and d for the recessive allele.

	Mother (dd) gametes	
	d	d
Father (Dd) gametes D		
Father (Dd) gametes d		

SOLVE

Now you have gathered the relevant information it is really important that you think carefully about how you need to use and present that information. Use the prompts below to help you.

To answer a **determine** question, you need to use the information you are given to work out the answer.

To help answer the question, let us recap some genetic terms. What is a dominant trait?

What is the difference between a dominant and a recessive trait?

What is the difference between genotype and phenotype?

In the question you are told to use D for the dominant allele and d for the recessive allele. Write down the genotypes of someone with polydactyly. There are two possible genotypes.

Now have a go at completing the Punnett square in the question. In each of the four empty squares, you need to write the letters of two alleles. One must come from the mother and one from the father.

From the genotypes you have written for the offspring, identify which genotype is heterozygous and which is homozygous.

The heterozygous genotype is: The homozygous genotype is:

Now identify the phenotype of each combination underneath the alleles in the Punnett square. In each of the four squares you should write 'polydactyly' or 'unaffected'.

Complete the sentences below to describe the probability of the child having polydactyly. You can write probability as a fraction, percentage or ratio.

The probability of the child having polydactyly is:

The probability of the child **not** having polydactyly is:

There are additional practice questions with the writing frames for 'Compare' and 'Determine' at the end of the book.

Session 11 Biology practicals

<table>
<tr><td rowspan="2">A student carried out an experiment into the effect of light intensity on the rate of photosynthesis.

The student's results are shown in the table.

Describe how the student could analyse their results.</td><td rowspan="2">Distance of lamp from pondweed (cm)</td><td colspan="3">Number of bubbles per minute</td></tr>
<tr><td>Test 1</td><td>Test 2</td><td>Test 3</td></tr>
<tr><td></td><td>10</td><td>99</td><td>121</td><td>124</td></tr>
<tr><td></td><td>15</td><td>41</td><td>50</td><td>54</td></tr>
<tr><td></td><td>20</td><td>20</td><td>30</td><td>32</td></tr>
<tr><td></td><td>25</td><td>11</td><td>17</td><td>16</td></tr>
<tr><td></td><td>30</td><td>8</td><td>13</td><td>14</td></tr>
</table>

THINK

This question asks you to **describe** the results of a photosynthesis investigation. This means you will need to write down the steps of how to analyse the results accurately.

This question is asking about the results of an experiment measuring the rate of photosynthesis when you change the light intensity. You will need to know a bit about photosynthesis in plants and how light is used in the process. Without looking at the knowledge organiser, what do you already know about the rate of photosynthesis and light? What kind of graph would you use to display the results? Scribble what you know here, particularly key words and the equipment you would use.

Now write down the information you will need for this specific question. Not everything you noted down before about photosynthesis may be relevant.

Which key idea cards can help? Deal them out next to your work and note extra information here.

Don't worry about any blanks. Now have a go at the questions on Adapt© for session 11 to help you think further about this topic.

MAKING A SLIDE

1. Add a few drops of water onto the slide using a pipette.
2. Place the sample flat on the surface of the slide using tweezers.
3. Add one or two drops of iodine solution to the sample from a dropping bottle to stain certain structures and allow them to be seen more clearly.
4. Using a mounted needle or seeker, lower the coverslip onto the slide carefully to make sure there are no air bubbles.

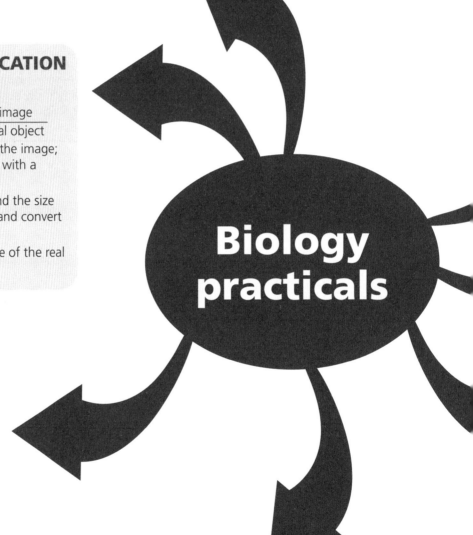

CALCULATING MAGNIFICATION OF AN IMAGE

$$\text{magnification of image} = \frac{\text{size of image}}{\text{size of real object}}$$

Sometimes you may be told the size of the image; sometimes you may need to measure it with a ruler yourself.

Make sure that the size of the image and the size of the real object are in the same unit, and convert if you need to.

Then divide the size of image by the size of the real object to calculate the magnification.

UNIT CONVERSION

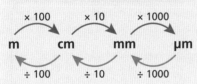

CALCULATING TOTAL MAGNIFICATION

Diagrams of magnified cells need to include the total magnification. If the eyepiece has a magnification of 10, which we write as ×10, and the objective lens has a magnification of ×40, then in total the cell has been magnified by ×400.

total magnification = eyepiece lens magnification × objective lens magnification

USING A MICROSCOPE

1. Select the lowest power objective lens.
2. Turn on the light and/or adjust the mirror so light is visible through the eyepiece lens.
3. Place the prepared slide onto the stage and secure with the stage clips.
4. Use the coarse focus wheel to bring the stage and objective lens as close to each other as possible without the lens touching the slide.
5. Look through the eyepiece and turn the coarse focus wheel so that the eyepiece lens and sample move apart, until the image comes into view.
6. Use the fine focus wheel to focus the image.

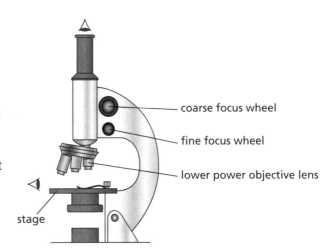

coarse focus wheel

fine focus wheel

lower power objective lens

stage

INVESTIGATING OSMOSIS

1. Cut plant tissue pieces to the same length and shape. Note down the starting mass for each piece.
2. Place the pieces of plant tissue into different solutions of salt.
3. Wait at least 10 minutes. Dry the pieces and then measure the mass of each piece of tissue.
4. Calculate the percentage change in mass:

$$\text{percentage change in mass} = \frac{(\text{final mass} - \text{starting mass})}{\text{starting mass}} \times 100$$

INVESTIGATING MOLECULES IN FOODS

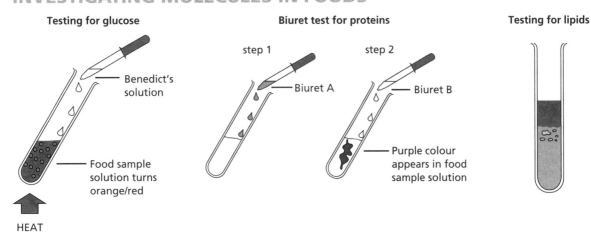

Testing for glucose

Benedict's solution

Food sample solution turns orange/red

HEAT

Biuret test for proteins

step 1

Biuret A

step 2

Biuret B

Purple colour appears in food sample solution

Testing for lipids

1. Make food solutions by grinding some food to a paste and adding some to a test tube.
2. To test for sugar: add Benedict's solution and then place in a water bath of boiling water.
3. A change to orange/red indicates that sugar is present.
4. To test for protein: first add Biuret A solution and then Biuret B solution.
5. A change to purple indicates that protein is present.
6. To test for lipids: filter some of the food solution into a test tube. Add Sudan III stain and shake gently.
7. A change to red/orange indicates that lipids are present.

INVESTIGATING RATE OF REACTION

The enzyme amylase breaks down starch. Iodine turns blue/black in the presence of starch.

1. Add a drop of iodine to each well.
2. In a test tube, mix starch and amylase.
3. Add a drop of starch/amylase mixture to the first well – this represents zero time.
4. After 30 seconds add another drop of the starch/amylase mixture to well 2.
5. Repeat every 30 seconds in wells 3, 4, etc., to show how long it takes for mixture to go from blue/black to brown.
6. Repeat steps 1–5 with different pH buffers to investigate the effects of pH.

drop of starch/amylase mixture added at zero time

spotting tile containing drops of iodine

INVESTIGATING PHOTOSYNTHESIS

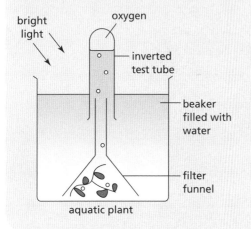

bright light

oxygen

inverted test tube

beaker filled with water

filter funnel

aquatic plant

1. Place an aquatic plant in a beaker of pond water under a funnel with an upturned test tube.
2. Place a lamp next to the beaker and measure how much oxygen from photosynthesis is collected in the test tube.
3. Move the lamp different distances from the plant – measure how the oxygen production changes.

INVESTIGATING REACTION TIME

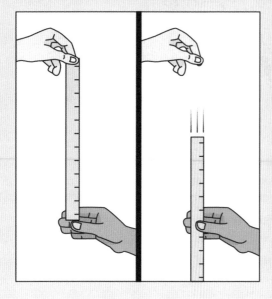

1. In a pair take it in turns for one student to drop a 30 cm ruler. The other student then catches the ruler between their thumb and forefinger.
2. Measure the distance the ruler drops.
3. Use a conversion table to convert distance of ruler drop into reaction time.
4. Repeat steps 1–3 to test the effects of a factor such as different levels of background noise.

MEASURING POPULATION SIZE

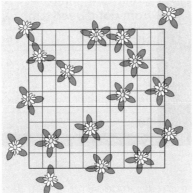

1. Choose which plant or plants you want to investigate.
2. Decide whether it is best to measure the number of plants or percentage cover.
3. Decide how to make your sample **random** and **representative**.
4. Place the **quadrat** down and count the presence of your plant or plants.
5. Keep repeating until your sample is representative.
6. Calculate the area covered by the plant:
   - Count each square that the plant covers in your quadrat.
   - Decide whether you will count only squares completely covered or also include partially covered squares.
   - To calculate the area of a quadrat multiply the height by the width (e.g. 100 cm × 100 cm = 10 000 cm^2 or 1 m^2.
   - To work out how many of your chosen plants there are in the habitat you are investigating: (1) divide the total area by the size of your quadrat; (2) multiply the average number of your plant per quadrat by the number of quadrats that fit in your habitat.

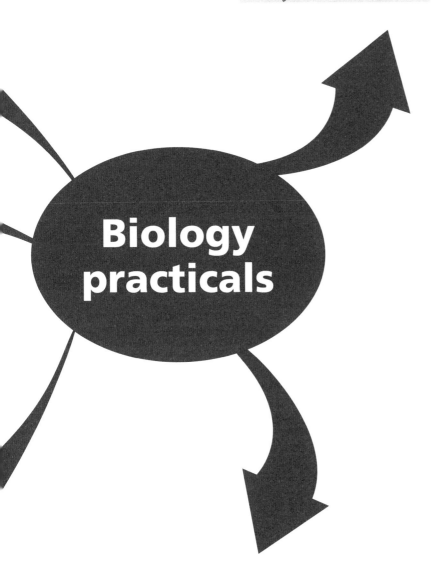

Biology practicals

ETHICS IN BIOLOGY EXPERIMENTS

1. If you are using people in your experiment, e.g. measuring reaction times, get informed consent before doing the experiment.
2. If you are using plant samples, e.g. investigating light, only pick a sustainable sample so that the plant can recover.
3. If you are measuring your sample in the environment, e.g. measuring population size, do not damage any plants or animals that you are observing.

Session 11   Solving the question

A student carried out an experiment into the effect of light intensity on the rate of photosynthesis.

The student's results are shown in the table.

Describe how the student could analyse their results.

Distance of lamp from pondweed (cm)	Number of bubbles per minute		
	Test 1	Test 2	Test 3
10	99	121	124
15	41	50	54
20	20	30	32
25	11	17	16
30	8	13	14

SOLVE

Now you've gathered the relevant information it is really important that you think carefully about how you need to use and present that information. Use the prompts below to help you.

What equipment would the student have used to measure the rate of photosynthesis?

To measure rate you need to measure how fast a product is made or how fast a reactant is used. Did the student measure a reactant or product in this investigation?

How did they change the light intensity?

Why does the rate of photosynthesis change when the light intensity changes?

From the results shown you can see the student carried out three repeats (tests) of the experiment. What calculation could you do to get a representative result for the bubbles counted at each distance?

Is the data repeatable or are there any anomalous results?

What kind of graph could they plot? What shape would the graph be?

There are additional practice questions with the writing frames for 'Design', 'Plan', 'Predict', 'Sketch' and 'Suggest' at the end of the book.

Session 12 The particle model

The table shows the density of some different materials.

a) Explain why large molecules generally have higher melting points than smaller molecules.

Use what you know about energy and changing state in your answer.

b) Explain why different materials have different densities.

Use data from the table and your knowledge of the particle theory to support your answer.

Substance	Density in kg/m³
Iron	8000
Gold	19000
Water	1000
Air	1.3

THINK

This question needs you to write in detail about the link between melting point and the size of a molecule. What do you already know about changing state, the energy needed to change state and the forces that hold molecules in a pattern? Without looking at the knowledge organiser, scribble some notes here.

Now write down the information you will need for this specific question. Not everything you noted down before about changing state may be relevant.

Which key idea cards can help? Deal them out next to your work and note extra information here.

Adapt *from Collins*

Don't worry about any blanks. Now have a go at the questions on Adapt© for session 12 to help you think further about this topic.

PARTICLE MODEL

Everything on Earth is made of matter that exists in three different states: solid, liquid, gas. The particle model states that matter consists of very tiny particles that are constantly moving. This model is not perfect, but it helps us to explain the properties of solids, liquids and gases. In reality, the gas particles are much further apart, and the particles are not all the same size.

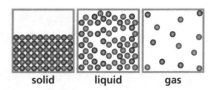

In solids, the particles are held together by strong forces of attraction, called bonds. A solid has a fixed size and shape. In liquids, the forces of attraction between the particles are weaker than in solids. There is no fixed structure. In gases the particles are almost free of any attractions between them. They move randomly and quickly, colliding with each other and the walls of their container.

CHANGING STATE

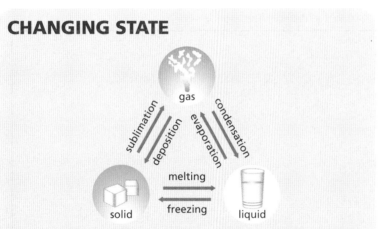

We can change the state of matter of any substance by heating or cooling. This will change the internal energy of a substance. As **thermal energy** is added to a solid, the particle vibrations increase and the forces holding the particles in place are stretched and pushed further apart. Eventually, some of the forces break and the particles can move over each other. The substance is now in the liquid state. This process is **melting**. When even more thermal energy is added, the remaining forces break and the particles are free to move from place to place. Changes of state are physical changes which differ from chemical changes because the material recovers its original properties if the change is reversed.

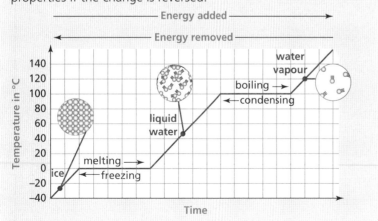

When you heat a substance there are stages where the heat added causes a rise in temperature. Then, as the substance changes state, the temperature stops rising because the heat is used to break the forces.

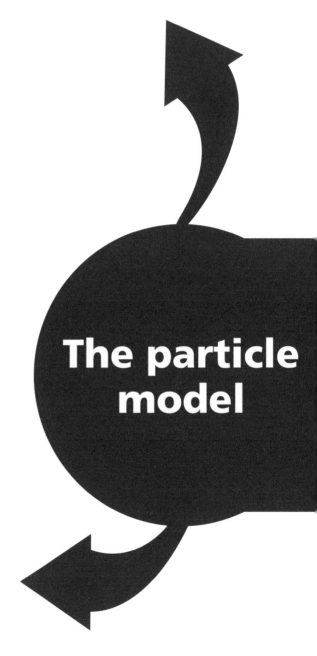

The particle model

DENSITY

Density compares the mass of materials that have the same volume. It is calculated from the equation

$$\text{density (kg/m}^3\text{)} = \frac{\text{mass (kg)}}{\text{volume (m}^3\text{)}}$$

The density of a material depends on the size of the particles and how closely they are packed. So most liquids are less dense than solids.

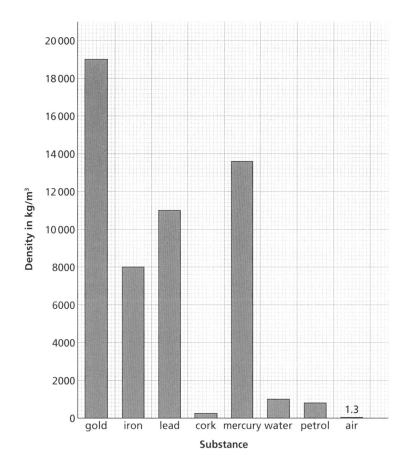

The **particle model** helps us to understand the differences between solids, liquids and gases. The **properties** of a material are determined by how particles are arranged. In a solid, particles can only vibrate in a fixed pattern. In liquids, particles can move over each other but are always touching other particles. Particles in a gas do not touch each other and there is lots of space between particles.

INTERNAL ENERGY

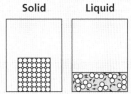

Particles are always moving. They have **kinetic energy**. At the same temperature, all particles have the same average kinetic energy. Therefore, heavy particles move more slowly than lighter particles.

Particles also have **potential energy**. The **internal energy** of a system is the total kinetic and potential energy of all the particles in the system. The amount of energy stored in or released from a system as the temperature changes can be calculated using the equation:

change in thermal energy = mass (kg) × specific heat capacity (J/kg °C) × temperature change (°C)

$$\Delta E = m\,c\,\Delta\theta$$

Session 12 Solving the question

The table shows the density of some different materials.

a) Explain why large molecules generally have higher melting points than smaller molecules.

Use what you know about energy and changing state in your answer.

b) Explain why different materials have different densities.

Use data from the table and your knowledge of the particle theory to support your answer.

Substance	Density in kg/m³
Iron	8000
Gold	19000
Water	1000
Air	1.3

SOLVE

Now you have gathered the relevant information it is really important that you think carefully about how you need to use and present that information. Use the prompts below to help you.

This question asks you to **explain** why more energy is needed for larger molecules to change state. You need to use what you know about energy changes that happen when a substance changes state. You should also use data from the table in your answer.

Draw the particles in solids, liquids and gases.

solid	liquid	gas

What is holding the particles in the patterns you have drawn?

For part a): The particles you have drawn represent molecules. With bigger molecules, do they have more or less force holding them together?

For part a): When a substance is heated, the energy makes the substance warmer. When the substance changes state, the energy is used to break the forces that hold the particles of the substance together. What happens to the temperature when a substance changes state?

For part a): Now bring together what you have written to explain why bigger molecules melt at higher temperatures.

For part b): Look at the data in the table. Write the state of each substance at room temperature (solid, liquid or gas).

iron: gold: water: air:

For part b): Write down the formula that links density with mass and volume. How does particle size affect the density of a substance?

For part b): Now bring together what you have written to explain why different materials have different densities.

There are additional practice questions with the writing frames for 'Determine', 'Evaluate' and 'Predict' at the end of the book.

Session 13 Atomic structure

The trends in physical and chemical properties in the periodic table depend on atomic structure and electronic structure.
Compare the reactivity of magnesium and calcium.

THINK

This question needs you to list how magnesium and calcium are the same and how they are different. What do you already know about the reactivity of magnesium and calcium? What about their atomic structure? Scribble down some ideas here.

Now write down the information you will need for this specific question. Not everything you noted down before about atomic structure and reactivity may be relevant.

Which key idea cards can help? Deal them out next to your work and note extra information here.

Adapt *from Collins*

Don't worry about any blanks. Now have a go at the questions on Adapt© for session 13 to help you think further about this topic.

ATOMIC STRUCTURE

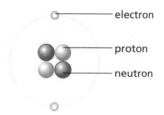

electron

proton

neutron

The nucleus of a helium atom contains two positively charged protons and two neutrons. It is surrounded by two negatively charged electrons. The relative masses and charges of the particles are shown in the table.

	Relative charge	Relative mass
Electron	−1	0.0005
Proton	+1	1
Neutron	0	1

The electrons in an atom are arranged in energy levels or shells. The first energy level can hold 2 electrons and the second and third levels can hold up to 8 electrons each. This atom of fluorine has 9 electrons: 2 in the first shell and 7 in the second shell. The **electronic structure** can be written 2,7 which makes it easy to find the element in the periodic table. This is because the group number is the same as the number of electrons in the outer shell of all the elements in that group.

FORMULA

All substances are chemicals made from atoms. You can see which elements are present by looking at the chemical formula. Elements from Group 1 and Group 7 combine to make compounds containing ions of each element in a fixed ratio of 1:1.

Ions with opposite charges are attracted together to form giant crystal lattices.

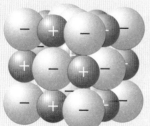

Group 1 element +
Group 7 elements −

The formula of sodium chloride is NaCl. The formula of magnesium chloride is $MgCl_2$ which tells us that there is one magnesium ion for every two chloride ions present.

oxygen molecule   water molecule   ethanoic acid molecule

The oxygen molecule contains 2 oxygen atoms and is written O_2. The water molecule contains 2 hydrogen atoms and 1 oxygen atom and is written H_2O. Ethanoic acid contains 2 carbon atoms, 2 oxygen atoms and 4 hydrogen atoms and is written CH_3CO_2H.

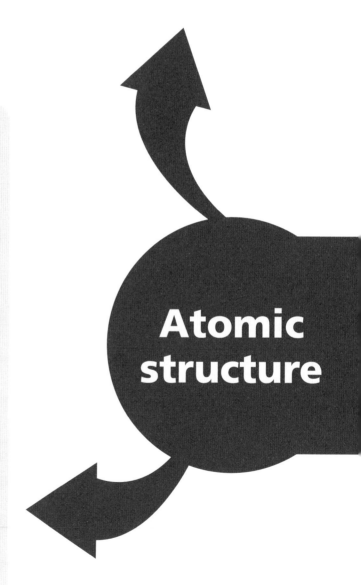

Atomic structure

PERIODIC TABLE

Group 1	2													Group 3	4	5	6	7	0
				$^{1}_{1}$H hydrogen															$^{4}_{2}$He helium
$^{7}_{3}$Li lithium	$^{9}_{4}$Be beryllium													$^{11}_{5}$B boron	$^{12}_{6}$C carbon	$^{14}_{7}$N nitrogen	$^{16}_{8}$O oxygen	$^{19}_{9}$F fluorine	$^{20}_{10}$Ne neon
$^{23}_{11}$Na sodium	$^{24}_{12}$Mg magnesium													$^{27}_{13}$Al aluminium	$^{28}_{14}$Si silicon	$^{31}_{15}$P phosphorus	$^{32}_{16}$S sulfur	$^{35}_{17}$Cl chlorine	$^{40}_{18}$Ar argon
$^{39}_{19}$K potassium	$^{40}_{20}$Ca calcium	$^{45}_{21}$Sc scandium	$^{48}_{22}$Ti titanium	$^{51}_{23}$V vanadium	$^{52}_{24}$Cr chromium	$^{55}_{25}$Mn manganese	$^{56}_{26}$Fe iron	$^{59}_{27}$Co cobalt	$^{59}_{28}$Ni nickel	$^{64}_{29}$Cu copper	$^{65}_{30}$Zn zinc	$^{70}_{31}$Ga gallium	$^{73}_{32}$Ge germanium	$^{75}_{33}$As arsenic	$^{79}_{34}$Se selenium	$^{80}_{35}$Br bromine	$^{84}_{36}$Kr krypton
$^{85}_{37}$Rb rubidium	$^{88}_{38}$Sr strontium	$^{89}_{39}$Y yttrium	$^{91}_{40}$Zr zirconium	$^{93}_{41}$Nb niobium	$^{96}_{42}$Mo molybdenum	$^{99}_{43}$Tc technetium	$^{101}_{44}$Ru ruthenium	$^{103}_{45}$Rh rhodium	$^{106}_{46}$Pd palladium	$^{108}_{47}$Ag silver	$^{112}_{48}$Cd cadmium	$^{115}_{49}$In indium	$^{119}_{50}$Sn tin	$^{122}_{51}$Sb antimony	$^{128}_{52}$Te tellurium	$^{127}_{53}$I iodine	$^{131}_{54}$Xe xenon

All known elements are listed in the periodic table (part of which is shown here). The metals are found on the left and the non-metals on the right. Each element has its own symbol, with two numbers next to it. If the symbol has more than one letter, then the first letter is always a capital letter. The **atomic number** (bottom number), tells us the number of protons in an atom. The number of electrons is the same as the number of protons. The **mass number** (upper number), tells us the total number of protons and neutrons present in an atom.

The elements are arranged in order of increasing atomic number. The table is divided into periods (rows) and groups (columns). Elements in the first period only have electrons in the first shell. Elements in the third period have electrons in the first three shells. The group number refers to the number of electrons in the outer shell, apart from Group 0, which contains the **noble gases**, where the outer electron shell is full, making them unreactive. The term Group 0 is used to avoid confusion because, although most elements in this group have 8 electrons in their outer shell, helium only has two.

Atoms are made of **neutrons, protons** and **electrons**. Each element has a specific number of protons in the **nucleus** and the same number of electrons in shells around the nucleus. The **periodic table** lists the number of protons of the elements and the symbol for each element. The periodic table is arranged so that you can see patterns in the types and strength of the reactions of each element relative to each other.

PERIODICITY

The trends in physical and chemical properties both across and down the periodic table depend on the atomic structure and the electronic structure. Melting and boiling points of elements in the same group change as you go down the group.

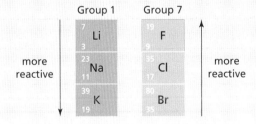

Groups contain elements with similar chemical properties. For example, the elements in Group 1 are all highly reactive because they have 1 electron in their outer shell. The reactivity increases down the group, as the outer electron is further away from the nucleus. Therefore, the **electrostatic attraction** with the nucleus is weaker. The outer electron is 'transferred' during chemical reactions. The reactivity of Group 7 increases as you go up the group because the number of electron shells decreases, making it easier for an electron to be attracted.

Session 13   Solving the question

> The trends in physical and chemical properties in the periodic table depend on atomic structure and electronic structure.
> Compare the reactivity of magnesium and calcium.

SOLVE

Now you have gathered the relevant information it is really important that you think carefully about how you need to use and present that information. Use the prompts below to help you.

This question asks you to **compare**. This means you need to write about the similarities and differences in the reactivity of magnesium and calcium.

Find magnesium and calcium in the periodic table. Which group are they in? What does this mean?

How many protons, neutrons and electrons does an atom of magnesium have? How many protons, neutrons and electrons does an atom of calcium have? Use the periodic table to find out. Write your answers in the table.

	No. of protons	No. of neutrons	No. of electrons
Magnesium			
Calcium			

Draw the electronic structure of calcium and magnesium. Use crosses for electrons. Label the energy levels.

Magnesium is above calcium in the periodic table. Which one is more reactive? Why?

Bring together your previous answers to make a list of how magnesium and calcium are similar.

Bring together your previous answers to make a list of how magnesium and calcium are different.

There are additional practice questions with the writing frames for 'Describe' and 'Determine' at the end of the book.

Session 14 Calculating chemical change

Magnesium (a metal) reacts with oxygen (a gas) to form a compound (a white powder).

a) Predict whether the solid product would be heavier, lighter or the same mass as the solid reactants.

Use an equation to explain your prediction.

b) Calculate how much magnesium oxide (MgO) would be made by the combustion of 5 g of magnesium (Mg).

THINK

This question needs you to make a decision about what happens to the atoms in the reaction. What can you remember about chemical reactions? Can you write an equation for this reaction already? Scribble down some ideas here.

Now write down the information you will need for this specific question. Not everything you noted down before about chemical reactions may be relevant.

Which key idea cards can help? Deal them out next to your work and note extra information here.

Adapt *from Collins*

Don't worry about any blanks. Now have a go at the questions on Adapt© for session 14 to help you think further about this topic.

CHEMICAL AND PHYSICAL CHANGE

During a chemical change, substances react together to make new substances with different properties.

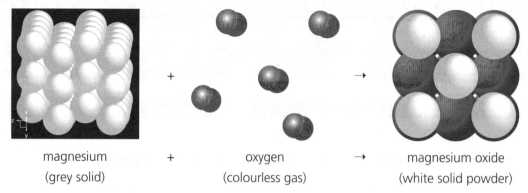

magnesium	+	oxygen	→	magnesium oxide
(grey solid)		(colourless gas)		(white solid powder)

When iodine **sublimes** it changes from a solid to a gas. A physical change takes place. The only change to the iodine atoms is an increase in internal energy.

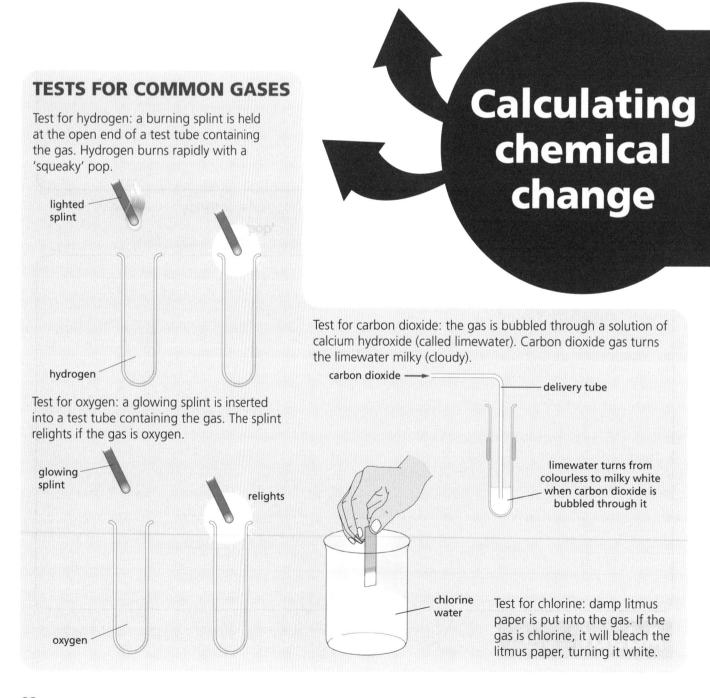

TESTS FOR COMMON GASES

Test for hydrogen: a burning splint is held at the open end of a test tube containing the gas. Hydrogen burns rapidly with a 'squeaky' pop.

lighted splint

'pop'

hydrogen

Test for oxygen: a glowing splint is inserted into a test tube containing the gas. The splint relights if the gas is oxygen.

glowing splint

relights

oxygen

Calculating chemical change

Test for carbon dioxide: the gas is bubbled through a solution of calcium hydroxide (called limewater). Carbon dioxide gas turns the limewater milky (cloudy).

carbon dioxide →

delivery tube

limewater turns from colourless to milky white when carbon dioxide is bubbled through it

chlorine water

Test for chlorine: damp litmus paper is put into the gas. If the gas is chlorine, it will bleach the litmus paper, turning it white.

WORD AND SYMBOL EQUATIONS

Chemical equations are used to describe what is happening in a chemical reaction.

reactants → products

For example:

zinc carbonate → zinc oxide + carbon dioxide

A word equation gives the names of the reactants (starting substances) and products (finishing products). The arrow tells us that a change is taking place.

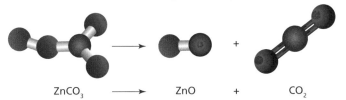

$ZnCO_3$ ⟶   ZnO   + CO_2

> During a chemical reaction, atoms are rearranged. All the atoms that were there at the beginning of the reaction still exist in the **products**. But the products will have different properties. Energy can be transferred from or to the surroundings during a chemical reaction. Chemical changes can be shown using word equations or symbol equations. As we know that no atoms are lost, we can calculate the amounts of **reactants** and products.

Symbol equations show the formulae of all the reactants and products. They should be balanced, showing mass is conserved during the reaction.

$Ca + O_2 → CaO$ ✗

This equation is not balanced. There are 2 oxygen atoms on the left side and only 1 on the right side. Each oxygen atom will join to a calcium atom. Therefore, we also need 2 calcium atoms on the left side.

$2Ca + O_2 → 2CaO$ ✓

It is now balanced.

CONSERVATION OF MASS

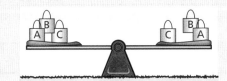

When a physical or chemical change takes place, no atoms are made or lost. During a chemical change, atoms are just rearranged. The masses of reactants and products are balanced. This is known as the law of **conservation of mass**. Energy can be transferred to or from the surroundings.

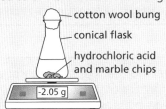

cotton wool bung
conical flask
hydrochloric acid and marble chips
-2.05 g

Some reactions may appear to involve a change in mass. This can usually be explained because a reactant or product is a gas and its mass has not been taken into account.

In this reaction

magnesium carbonate + hydrochloric acid → magnesium chloride + water + carbon dioxide

the mass appears to decrease. This is because carbon dioxide gas is produced and escapes from the flask into the atmosphere.

CALCULATING CHEMICAL QUANTITIES

In a reaction, how much product will we make? How much reactant is needed? We can calculate the amounts of reactants and products in a chemical reaction using the balanced equation and the **relative formula mass** (M_r).

M_r of a compound is the sum of the **relative atomic masses** (A_r) of the atoms in the numbers shown in the formula. You can take the A_r value as the mass number from the periodic table.

A_r: Zn = 65; C = 12; O = 16

$ZnCO_3$	→	**ZnO**	+ **CO_2**
M_r: 65 + 12 + (16 × 3)		65 + 16	+ 12 + (16 × 2)
65 + 12 + 48	→	81	+ 12 + 32
125	→	81	+ 44
125	→		125

If we start with 12.5 g of $ZnCO_3$, the multiplier is $\frac{12.5}{125} = 0.1$

We will produce 0.1 × 81 = 8.1 g of ZnO and 0.1 × 44 = 4.4 g of CO_2.

The total mass of the reactants equals the total mass of the products.

Use a multiplier to scale the amounts up or down.

Session 14   Solving the question

Magnesium (a metal) reacts with oxygen (a gas) to form a compound (a white powder).
a) Predict whether the solid product would be heavier, lighter or the same mass as the solid reactants.

   Use an equation to explain your prediction.
b) Calculate how much magnesium oxide (MgO) would be made by the combustion of 5 g of magnesium (Mg).

SOLVE

Now you have gathered the relevant information it is really important that you think carefully about how you need to use and present that information. Use the prompts below to help you.

This question asks you to **predict** the likely result of magnesium reacting with oxygen. You need to use what you know about what happens to atoms in reactions. You also need to **calculate** the amount of magnesium oxide formed.

For part a): Decide what the reactants and products of the reaction are. Write a word equation.

For part a): Now write a formula equation.

For part a): Is the formula equation balanced? In other words, are there the same number of atoms of each element on both sides of the equation? If not, balance it.

For part a): Now look at the equation again. Which reactants and which products are solids?

For part a): Now look at the balanced equation to compare how many atoms of the solid Mg (2) react to form how many molecules of the solid MgO (2). You can now predict whether the mass of product (2MgO) is heavier, lighter or the same mass as the magnesium reacted (2Mg).

For part b): The relative atomic mass of Mg is 24 and O is 16. Calculate the relative formula mass of MgO.

For part b): Using the equation $2Mg + O_2 \rightarrow 2MgO$, if we start with 24 g of Mg, how much MgO will be made?

For part b): When we start with 5 g of magnesium, the mass of magnesium oxide produced is scaled down by the same proportion. Work out the multiplier.

For part b): Now bring together your previous answers to work out how much MgO is formed from 5 g of Mg.

There are additional practice questions with the writing frames for 'Balance', 'Calculate', 'Define' and 'Justify' at the end of the book.

Session 15 Bonding

The table gives some information about the atoms of sodium, oxygen and chlorine.

Element	Atomic number	Relative atomic mass	State at room temperature
Oxygen	8	16	Gas
Sodium	11	23	Solid
Chlorine	17	35.5	Gas
Sodium oxide			Solid

a) Suggest how sodium oxide crystals and molecules of chlorine gas are formed.

b) Explain why sodium oxide is a solid but chlorine is a gas.

Use what you know about ionic bonding and covalent bonding in your answer.

THINK

This question needs you to remember how ionic and covalent compounds are formed. What can you remember about the electron arrangements of atoms? Can you draw sodium, oxygen and chlorine atoms? What about their ions? Scribble down some ideas here.

Now write down the information you will need for this specific question. Not everything you noted down before about ionic and covalent compounds may be relevant.

Which key idea cards can help? Deal them out next to your work and note extra information here.

Don't worry about any blanks. Now have a go at the questions on Adapt© for session 15 to help you think further about this topic.

IONIC BONDING

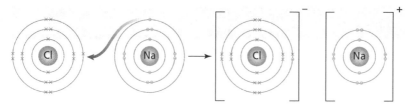

When a metal atom reacts with a non-metal atom, electrons in the outer shell of the metal atom are transferred. Both positively charged and negatively charged ions are formed. The ions are held together by an electrostatic force of attraction called the ionic bond. The electron transfer during the formation of an **ionic compound** can be represented by a **dot and cross diagram**. The dots represent electrons from one atom and the crosses represent electrons from another atom.

The ions produced have the electronic structure of a noble gas. Sometimes you might see the dot and cross diagram written like this:

Metals in Group 1 and Group 2 of the periodic table produce ions with a charge of + and 2+, respectively. Non-metals in Group 6 and Group 7 produce ions with a **charge** of 2– and –, respectively. Ionic compounds form **giant crystal lattices**.

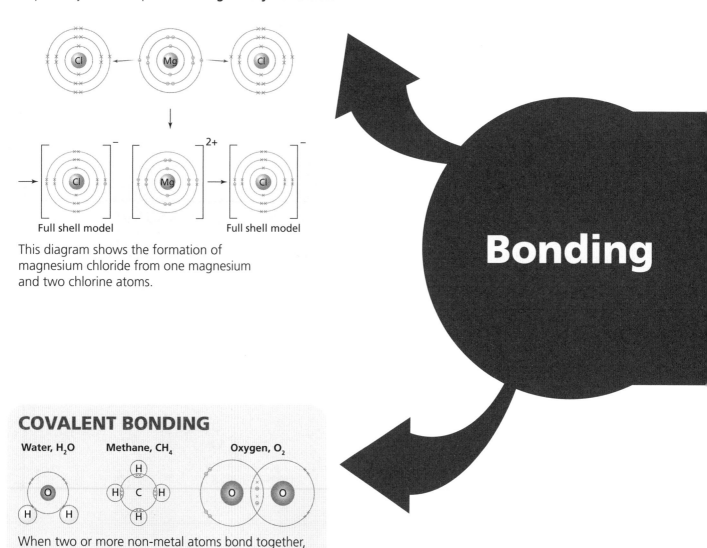

Full shell model   Full shell model

This diagram shows the formation of magnesium chloride from one magnesium and two chlorine atoms.

Bonding

COVALENT BONDING

Water, H_2O Methane, CH_4 Oxygen, O_2

When two or more non-metal atoms bond together, they form a molecule. Covalent molecules are held together by covalent bonds, where electron pairs are shared between two atoms. Dot and cross diagrams are used to represent covalent bonding.

TYPE OF BONDING

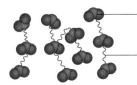

strong covalent bonds between the atoms

intermolecular forces between the molecules

A chemical bond is a force of attraction between different particles. There are two types of bonds found in and between simple molecules such as water (H_2O). A strong covalent bond holds the hydrogen and oxygen atoms together in each molecule. Weaker bonds called **intermolecular forces** exist between the water molecules.

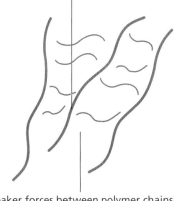

strong covalent bonds

weaker forces between polymer chains

Intermolecular forces play an important role in **polymers**, explaining why polymers can be stretched without breaking. The atoms in the polymer molecules are linked to other atoms by strong covalent bonds to form a long chain. The intermolecular forces between the polymer chains are weaker. They allow the polymer chains to slide over each other but are not so weak to allow the chains to be pulled apart.

Electrons can be shared with another atom in a **covalent bond**. Electrons can also move between atoms to form ions during the formation of an **ionic bond**. In both cases, atoms share, donate or accept an electron to gain a full outer shell. In metals, the metal ions are surrounded by a sea of electrons donated by the metal atoms. How particles are arranged depends on the strength of the bonds and forces that hold the particles in position.

METALLIC BONDING

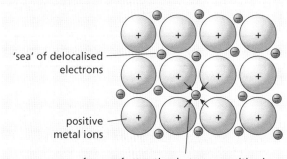

'sea' of delocalised electrons

positive metal ions

forces of attraction between positive ions and negative electrons pull ions together

Metals consist of giant structures of atoms arranged in a regular pattern. Each metal atom has one or more electrons in its outer shell, which are free to move between and around other atoms. These electrons are called **delocalised electrons**. The metal atoms that 'lose' these delocalised electrons become positively charged ions. The **metallic bond** is the strong electrostatic force of attraction between the positive metal ions and the delocalised electrons.

Session 15   Solving the question

The table gives some information about the atoms of sodium, oxygen and chlorine.

Element	Atomic number	Relative atomic mass	State at room temperature
Oxygen	8	16	Gas
Sodium	11	23	Solid
Chlorine	17	35.5	Gas
Sodium oxide			Solid

a) Suggest how sodium oxide crystals and molecules of chlorine gas are formed.

b) Explain why sodium oxide is a solid but chlorine is a gas.

Use what you know about ionic bonding and covalent bonding in your answer.

SOLVE

Now you have gathered the relevant information it is really important that you think carefully about how you need to use and present that information. Use the prompts below to help you.

This question asks you to **suggest** how a crystal of sodium oxide and how a molecule of chlorine is formed. To answer this question you need to apply what you know about ionic bonding and covalent bonding. You then need to **explain**. Use what you know to write why sodium oxide is a solid but chlorine is a gas.

For part a): Ionic bonds form when electrons are lost by one atom and given to another. What happens to the charge of an atom when an electron is lost? What happens to the charge of an atom when an electron is gained?

For part a): How many electrons has each sodium atom lost?

For part a): How many electrons has each oxygen atom gained?

For part a): How many sodium ions are needed for each oxygen atom? Write this number next to the sodium in the formula for sodium oxide. Make sure the N of sodium is capitalised and the a is lower-case. The O of oxygen is a capital.

For part a): Now write a sentence about what happens when an ionic bond is formed between sodium and oxygen.

For part a): Now bring together your ideas to write a sentence about what happens when chlorine atoms form covalent bonds to make a chlorine molecule.

For part b): Now bring together what you have written and try to explain why sodium oxide is a solid but chlorine is a gas.

There are additional practice questions with the writing frames for 'Draw' at the end of the book.

Session 16 Properties of materials

Graphite has two helpful properties: It conducts electricity and it is used as a solid lubricant.

a) Explain why graphite has both properties.

b) Diamond is another form of carbon. Explain why diamond is used for drill tips whereas graphite is not.

Use what you know about how the carbon is bonded in each substance in your answer.

THINK

This question needs you to link the structures of graphite and diamond to what they are used for. What can you remember about atoms of carbon? What can you remember about covalent bonding? Can you draw a molecule of graphite? Scribble down some ideas here.

Now write down the information you will need for this specific question. Not everything you noted down before about graphite and diamond may be relevant.

Which key idea cards can help? Deal them out next to your work and note extra information here.

Adapt *from Collins*

Don't worry about any blanks. Now have a go at the questions on Adapt© for session 16 to help you think further about this topic.

SMALL MOLECULES

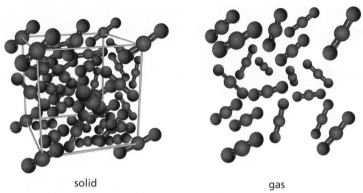

solid   gas

Substances with small covalent molecules have low melting and boiling points, so usually exist as liquids or gases at room temperature. Solids are formed when cooled to lower temperatures, e.g. carbon dioxide. The weak intermolecular forces between the molecules mean they are easy to separate. They do not conduct electricity as there are no delocalised electrons or ions available to move.

GIANT COVALENT STRUCTURES

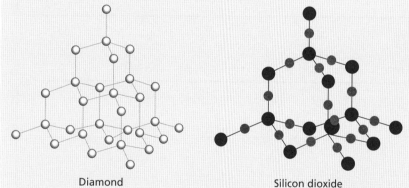

Diamond   Silicon dioxide

Substances with **Giant covalent molecules** such as diamond (C) and sand (SiO_2) have high melting and boiling points. It takes a lot of energy to break the strong covalent bonds formed between each atom. These materials are very hard and do not conduct electricity as there are no delocalised electrons or ions available.

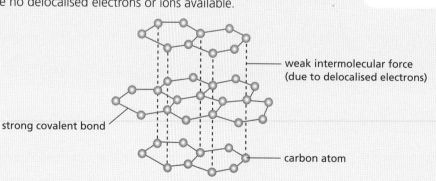

weak intermolecular force
(due to delocalised electrons)

strong covalent bond

carbon atom

Graphite is also a giant covalent molecule but it is soft and does conduct electricity. It does not follow the rule. This is because, in graphite, carbon atoms form hexagonal rings of carbon atoms in layers. It also has free or delocalised electrons that can move through the structure. It is the free electrons that allow graphite to conduct electricity.

Properties of materials

IONIC COMPOUNDS

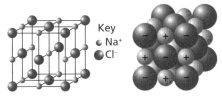

Key
- Na⁺
- Cl⁻

Ionic compounds such as sodium chloride form giant crystal lattices. We can use different types of 3D models to represent the structures, such as a ball and stick model or a close packed model.

These models are useful but they have limitations. The ball and stick model is limited because the attractive force between the ions is shown as a solid stick. The close packed model is limited as it shows the ions as solid spheres, when in reality a lot of the ion is empty space.

Ionic compounds have high melting and boiling points because the forces of attraction holding the ions together are very strong. They conduct electricity when melted or dissolved in water because the ions are free to move and so charge can flow.

The melting point of magnesium oxide, MgO, is 2852 °C. This is explained by its ionic bonding.

(a) dot and cross diagram

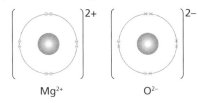

Mg^{2+} O^{2-}

(b) 3D close packed diagram

Remember that it takes a lot of energy to break the strong forces of attraction between the 2+ and 2− ions.

The properties of a material depend on how the atoms are bonded, the size of the molecules and the arrangement of the atoms and ions. Small molecules have weak forces between them so they have low boiling points and are normally gases. Metals are good conductors of electricity because of their sea of delocalised electrons.

METALS AND ALLOYS

The delocalised electrons in metals make them good conductors of electricity. The layers of atoms can easily slide over each other which means they can be bent and shaped. They are **malleable** (can be hammered into shape) and **ductile** (can be pulled into wires). They have high melting and boiling points due to the strong metallic bond.

An **alloy** is a mixture. The different sized atoms twist or distort the layers, making it more difficult for them to slide over each other. This means that alloys are harder than pure metals. For example, steel is an alloy made from iron mixed with carbon. It is used in the construction industry because it is very strong.

Metal	Lead	Tin	Solder
Melting point in °C	327	232	183

Solder is an alloy made from lead and tin. It can be used to join two metals together, due to its low melting point.

Session 16   Solving the question

Graphite has two helpful properties: it conducts electricity and it is used as a solid lubricant.
a) Explain why graphite has both properties.
b) Diamond is another form of carbon. Explain why diamond is used for drill tips whereas graphite is not.
Use what you know about how the carbon is bonded in each substance in your answer.

SOLVE

Now you have gathered the relevant information it is really important that you think carefully about how you need to use and present that information. Use the prompts below to help you.

This question asks you to **explain**. You need to use what you know about the structures of graphite and diamond to write why they have the uses mentioned.

For part a): Carbon is in Group 4 of the periodic table. How many electrons does it have available for bonding in its outer shell?

For part a): In graphite, each carbon atom only uses three of its available electrons for bonding, and has one free electron left over.

The diagram below shows the structure of graphite. Add labels to the diagram to show the types of bond and the atoms present. What is the name given to the free electrons?

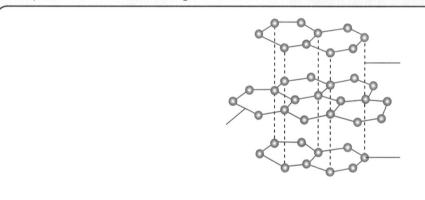

For part a): Which of the features you have labelled makes graphite a solid lubricant? Explain this.

For part a): Which of the features you have labelled makes graphite a conductor of electricity? Explain this.

For part b): What physical property does a material need if it is going to be used as a drill tip?

For part b): Suggest a reason why graphite is soft.

For part b): Diamond forms a giant molecular structure. Why is diamond harder than graphite?

There are additional practice questions with the writing frames for 'Evaluate' and 'Justify' at the end of the book.

Session 17 Elements, compounds and mixtures

Chromatography is a technique used to separate mixtures of different compounds or to test a compound for purity.

The diagram shows a chromatogram.

a) Describe how chromatography is used to identify the dyes used in food colouring.

b) Calculate the R_f value of the spot labelled S on the diagram.

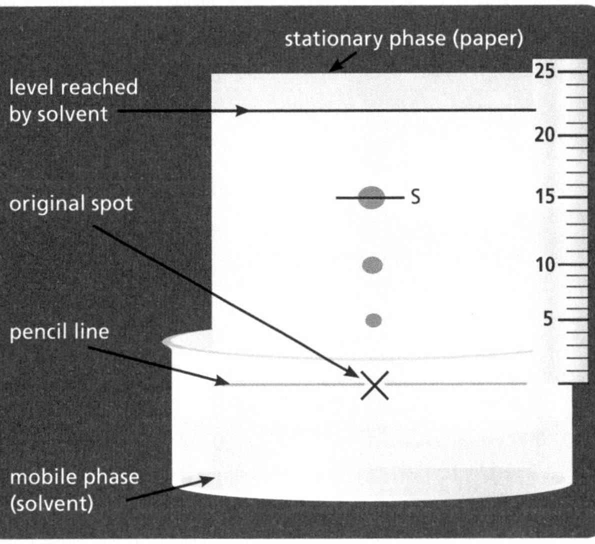

stationary phase (paper)
level reached by solvent
original spot
pencil line
mobile phase (solvent)
S
25
20
15
10
5

THINK

This question needs you to write about chromatography. What can you remember about separating mixtures and compounds? Scribble down some ideas here.

Now write down the information you will need for this specific question. Not everything you noted down before about separation techniques may be relevant.

Which key idea cards can help? Deal them out next to your work and note extra information here.

Adapt *from Collins*

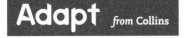

Don't worry about any blanks. Now have a go at the questions on Adapt© for session 17 to help you think further about this topic.

ELEMENTS, COMPOUNDS AND MIXTURES

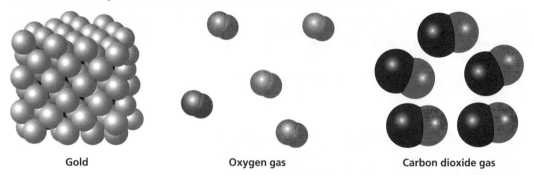

Gold   **Oxygen gas**   **Carbon dioxide gas**

Gold is an element. It only contains atoms of gold. Oxygen gas is an element. It only contains oxygen atoms in its molecules. Carbon monoxide is a compound. Its molecules consist of one atom of carbon chemically bonded to one atom of oxygen. Air is a mixture of oxygen, nitrogen, water vapour and other gases. The chemical properties of each gas in the air remain unchanged. We can separate the gases in the air using **distillation**.

SEPARATION TECHNIQUES

Mixtures can be separated by physical processes that do not involve chemical reactions. For example, we can use a magnet to separate iron, cobalt or nickel from a pile of scrap metals such as copper and aluminium.

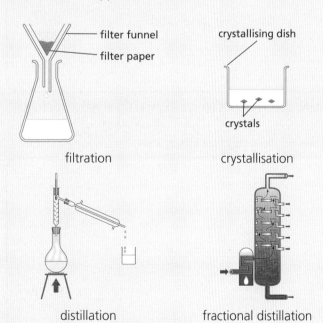

filtration   crystallisation

distillation   fractional distillation

Elements, compounds and mixtures

Filtration is used to separate **insoluble** substances like sand and mud from a **soluble** substance such as sodium chloride. Put the mixture in a beaker with some water. Then pour it into the filter. The insoluble part (sand) will collect on the filter paper. The soluble part (salt solution) will collect in the flask. **Crystallisation** is used to remove water from a solution, leaving solid **crystals** behind. This is done by gently heating the solution so the water evaporates. Distillation is the process of evaporation, followed by condensation. It is used to separate mixtures of liquids with different boiling points, for example alcohol and water. **Fractional distillation** is used to separate **crude oil**. The mixture of liquids is vaporised and compounds with different boiling points condense at different temperatures.

CHROMATOGRAPHY

Carrying out chromatography

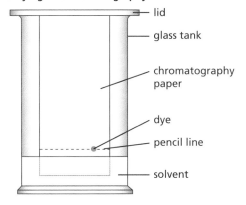

- lid
- glass tank
- chromatography paper
- dye
- pencil line
- solvent

Chromatogram of dyes of food colours

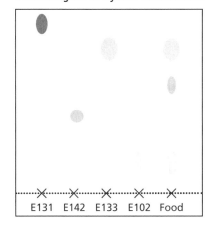

E131   E142   E133   E102   Food

Calculating R_f values

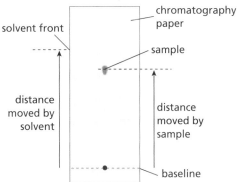

- solvent front
- chromatography paper
- sample
- distance moved by solvent
- distance moved by sample
- baseline

Elements are substances made out of only one type of atom. **Mixtures** are made out of two or more substances that are not chemically bonded together. **Compounds** are substances that contain at least two different elements chemically bonded together. The parts of a compound cannot be easily separated. The parts of a mixture can be separated using the different properties of each part of the mixture. For example, the different colours of ink in a pen can be separated because they have different solubilities.

Chromatography uses the property of solubility to separate mixtures of inks, dyes, food colouring or amino acids. During the process, a **solvent** moves through the paper carrying different compounds over different distances. The distance a compound moves depends on how soluble it is in the solvent. The end result is a **chromatogram**, which can be used to identify different parts of the mixture. This is done by calculating and comparing R_f values.

$$R_f = \frac{\text{distance moved by substance}}{\text{distance moved by solvent}}$$

PURE AND IMPURE SUBSTANCES

 A pure compound

 A mixture of compounds

Melting point and boiling point data can be used to tell apart pure substances from mixtures. If a substance is iempure the melting point will be *lower* than for the pure substance and the substance will melt over a *broad* range of temperatures. If a substance is *pure* it will melt more *sharply* at a specific temperature. If a substance is *impure* the boiling point will be *higher* than the pure substance.

Session 17 Solving the question

Chromatography is a technique used to separate mixtures of different compounds or to test a compound for purity.

The diagram shows a chromatogram.

a) Describe how chromatography is used to identify the dyes used in food colouring.

b) Calculate the R_f value of the spot labelled S on the diagram.

stationary phase (paper)

level reached by solvent

original spot

pencil line

mobile phase (solvent)

S

25 20 15 10 5

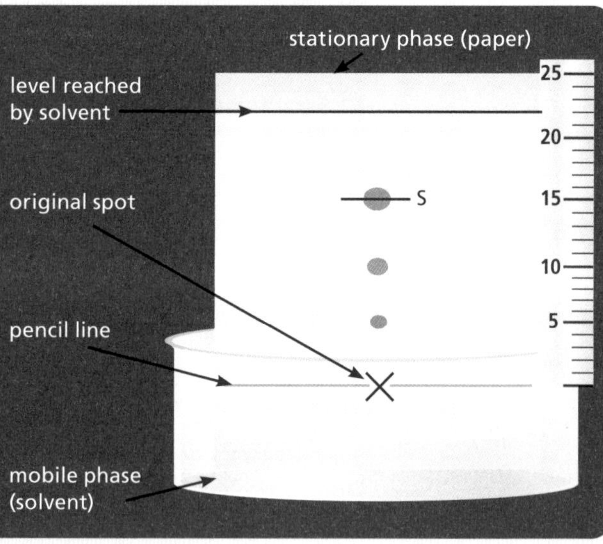

SOLVE

Now you have gathered the relevant information it is really important that you think carefully about how you need to use and present that information. Use the prompts below to help you.

This question asks you to **describe**. This means you need to communicate how chromatography works. You then need to **calculate** a value. You will need to recall a formula and show your working.

For part a): Chromatography is used to separate mixtures. What is a mixture?

For part a): Which physical property is used in chromatography to do the separation?

For part a): What is meant by the terms mobile phase and stationary phase?

For part a): Why is a pencil used to draw the starting line?

For part a): Look at the diagram showing the chromatogram. How many dyes were found in the sample?

For part a): We can identify the dyes by calculating the R_f value. Write down the equation that links R_f values to the distance travelled by the spot.

For part a): Now bring together these ideas to describe how chromatography is used to identify the dyes used in food colouring.

For part b): Calculate the R_f value of the spot labelled S on the diagram.

There are additional practice questions with the writing frames for 'Justify' at the end of the book.

Session 18 Earth chemistry

Earth's atmosphere was formed from chemical reactions and biological processes taking place over billions of years.

Crude oil is a fossil fuel formed millions of years ago.

Compare how photosynthesis and burning fossil fuels have changed the Earth's atmosphere.

THINK

This question needs you to know about the processes of photosynthesis and combustion. What can you remember about photosynthesis? What do you know about burning? Scribble down some ideas here.

Now write down the information you will need for this specific question. Not everything you noted down before about photosynthesis and combustion may be relevant.

Which key idea cards can help? Deal them out next to your work and note extra information here.

Adapt *from Collins*

Don't worry about any blanks. Now have a go at the questions on Adapt© for session 18 to help you think further about this topic.

HYDROCARBONS

Crude oil is a mixture of hydrocarbon molecules. A hydrocarbon contains only carbon and hydrogen atoms. Most hydrocarbons in crude oil are **alkanes**. Alkanes have the general formula C_nH_{2n+2} where n is the number of carbon atoms. The first four alkanes are methane, ethane, propane and butane.

Name	Number of carbon atoms	Formula	Displayed formula
Methane	1	CH_4	H—C—H with H above and H below
Ethane	2	C_2H_6	H—C—C—H with H's above and below

When hydrocarbons burn in oxygen, carbon dioxide and water are produced. Energy is given out to the surroundings in a **combustion** reaction. Fuels extracted from crude oil include petrol, diesel and methane gas which are used for heating and transport.

The equation for the combustion of methane is

methane   +   oxygen   →   carbon dioxide   +   water

CH_4 + $2O_2$ → CO_2 + $2H_2O$

Earth chemistry

FRACTIONAL DISTILLATION

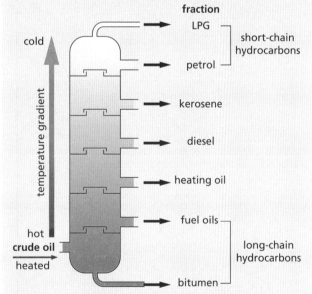

Fractional distillation is used to separate crude oil. Crude oil is heated until it forms a **vapour**. The vapour rises up the tower. The different hydrocarbons **condense** at the place in the column where the temperature is just below their **boiling point**. Short-chain hydrocarbons have lower boiling points than long-chain hydrocarbons so rise further up the tower. The different fractions are then collected. A fraction is a group of hydrocarbon molecules with similar boiling points.

USEFUL MATERIALS

It is not just fuels that come from crude oil. Different fractions are processed to make plastics, detergents, solvents and bitumen for roads and roofing. Crude oil is used as a raw material in many different industries.

EARTH'S ATMOSPHERE

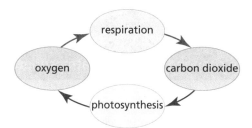

Air is a mixture of different gases. The early atmosphere contained gases such as ammonia and carbon dioxide from volcanoes. Water vapour condensed to form oceans. Algae used carbon dioxide for photosynthesis and started to produce oxygen. Carbon dioxide dissolved in the oceans.

argon 1%, carbon dioxide 0.04%
oxygen approx. 20%
nitrogen approx. 80%

The proportions of gases in today's atmosphere remain almost constant, apart from water vapour. Plants photosynthesise, releasing oxygen.

carbon dioxide + water → glucose + oxygen

Animals and plants respire, releasing carbon dioxide.

glucose + oxygen → carbon dioxide + water

Earth's atmosphere was formed from chemical reactions and biological processes taking place over millions of years. Crude oil is a fossil fuel made from the remains of organisms that lived millions of years ago. It is a mixture of **hydrocarbons** that can be made into useful substances via physical and chemical processes. The hydrocarbons found in crude oil can be separated into fractions in a process called fractional distillation. Chemical reactions are used to make substances such as **polymers**, plastics and detergents, all with different properties.

POLYMERS

Polymers are very large molecules, formed from many smaller molecules called **monomers** in a process called **polymerisation**. Many polymers such as **poly(ethene)** are made from **alkenes**, another hydrocarbon. Alkenes are produced by '**cracking**' some of the larger less useful alkanes produced during fractional distillation of crude oil.

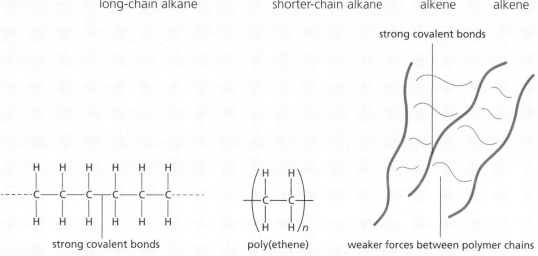

$C_{16}H_{34}$ →(cracking) C_8H_{18} + $2C_3H_6$ + C_2H_4
long-chain alkane — shorter-chain alkane — alkene — alkene

strong covalent bonds

strong covalent bonds

poly(ethene)

weaker forces between polymer chains

The atoms in polymer molecules are linked to other atoms by strong covalent bonds. Weaker intermolecular forces between the polymer chains hold the chains together. Polymers are usually stretchy solids as the chains can slide over each other when a force is applied.

79

Session 18   Solving the question

Earth's atmosphere was formed from chemical reactions and biological processes taking place over billions of years.
Crude oil is a fossil fuel formed millions of years ago.
Compare how photosynthesis and burning fossil fuels have changed the Earth's atmosphere.

SOLVE

Now you have gathered the relevant information it is really important that you think carefully about how you need to use and present that information. Use the prompts below to help you.

This question asks you to **compare**. This means you need to describe the similarities and differences between how burning fossil fuels and photosynthesis have changed the Earth's atmosphere.

What is photosynthesis? Describe photosynthesis in words or write the equation.

What is crude oil made from and where was it made?

Crude oil is a mixture of different substances. Describe how fractional distillation is used to separate them.

What is combustion? Describe combustion in words or write the equation.

Now think about the development of the Earth's atmosphere. When did photosynthesis start happening? When did combustion start happening?

Now bring together your answers to compare the two processes. You might like to think about what each process added or removed from the atmosphere, and when each one happened.

There are additional practice questions with the writing frames for 'Suggest' at the end of the book.

Session 19 Electrolysis

a) Describe how pure aluminium is extracted from aluminium ore.

b) Explain why aluminium is so expensive.

THINK

For this question you need to know about the extraction of a metal and the process of electrolysis. Can you remember how electrolysis works? What do you know about how aluminium is found naturally and how it is extracted? Scribble down some ideas here.

Now write down the information you will need for this specific question. Not everything you noted down before about electrolysis in industry may be relevant.

Which key idea cards can help? Deal them out next to your work and note extra information here.

Adapt *from Collins*

Don't worry about any blanks. Now have a go at the questions on Adapt© for session 19 to help you think further about this topic.

EXTRACTING METALS

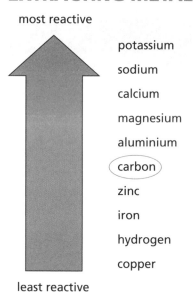

most reactive

potassium
sodium
calcium
magnesium
aluminium
carbon
zinc
iron
hydrogen
copper

least reactive

Unreactive metals like gold are found in the earth. Most metals need to be extracted from their ores by chemical reactions. The method of extraction depends on the reactivity.

Metals that are less reactive than carbon are extracted using carbon. The metal oxide is **reduced** as carbon removes the oxygen.

zinc oxide + carbon → zinc + carbon dioxide

Metals more reactive than carbon are extracted by electrolysis. Large amounts of energy are used in the process to melt the compounds and to produce the electricity needed for electrolysis. This means metals such as aluminium are expensive.

ELECTROLYSIS

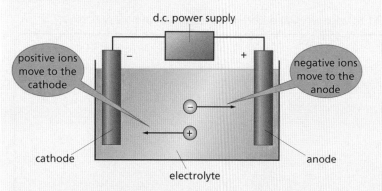

d.c. power supply

positive ions move to the cathode

negative ions move to the anode

cathode

anode

electrolyte

During electrolysis direct current is passed through a molten ionic compound or a solution of an ionic compound, called the **electrolyte**. The ions are attracted towards the electrodes where they are **discharged** (they either gain or lose electrons), producing elements.

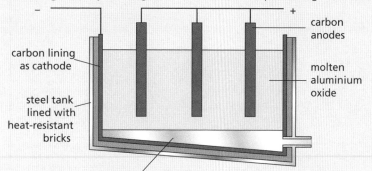

carbon anodes

carbon lining as cathode

molten aluminium oxide

steel tank lined with heat-resistant bricks

molten aluminium collects at the bottom

Aluminium is manufactured by the electrolysis of a molten mixture of aluminium oxide and cryolite using carbon electrodes. The cryolite is added to reduce energy costs by lowering the melting temperature of the aluminium oxide. Aluminium is produced at the cathode (– electrode) and oxygen at the anode (+ electrode). The positive electrode (carbon anode) must be continually replaced because the oxygen reacts with it producing carbon dioxide gas.

Electrolysis

CONSERVATION OF MASS

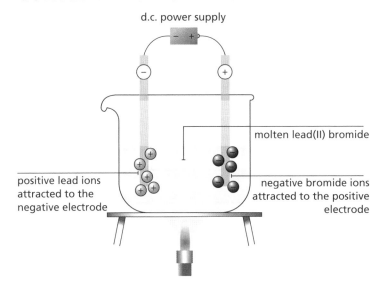

d.c. power supply

molten lead(II) bromide

positive lead ions attracted to the negative electrode

negative bromide ions attracted to the positive electrode

Solid lead bromide ($PbBr_2$) is a crystal lattice made from Pb^{2+} and Br^- ions. It does not conduct electricity because the ions cannot move. When solid $PbBr_2$ is heated to its melting point, the strong electrostatic forces between the Pb^{2+} and Br^- ions are broken. $PbBr_2$ becomes a liquid. The ions can now move. When a direct electric current is passed through the liquid, the ions will move towards the electrodes as it conducts electricity.

Pb^{2+} ions go to the negative electrode and lead (Pb) is formed.

Br^- ions go to the positive electrode, and bromine (Br_2) is formed.

No atoms are lost during electrolysis. Mass is conserved.

During a chemical reaction, atoms are rearranged. All the atoms that were there at the beginning of the reaction still exist in the products. Chemical changes can be communicated using word and formula equations. Ionic compounds can be rearranged by **electrolysis**. In electrolysis positive ions are attracted to the negative **cathode** and negative ions are attracted to the positive **anode**. This is a useful process for extracting pure reactive metals such as aluminium from its **ore**.

ELECTROLYSIS OF AQUEOUS SOLUTIONS

The electrolysis of aqueous solutions is more complicated. As well as ions from the salt there are also hydrogen ions (H^+) and **hydroxide ions** (OH^-) present from the water. This means two ions are attracted to each electrode but only one ion is discharged. The most reactive ion stays in solution and the least reactive ion is discharged. The table summarises which ions are discharged.

At the cathode	At the anode
If the metal ion is reactive, then hydrogen is formed	If a sulfate group is present, then oxygen is formed
If the metal ion is unreactive, then the metal is formed	If a halide ion is present then the halogen is formed

For example, in a solution of copper sulfate the ions are: Cu^{2+}, H^+, SO_4^{2-}, OH^-.

Copper is formed at the negative electrode and oxygen is formed at the anode.

In a solution of sodium chloride the ions present are: Na^+, H^+, Cl^-, OH^-.

Hydrogen is formed at the cathode and chlorine is formed at the anode.

Session 19 Solving the question

a) Describe how pure aluminium is extracted from aluminium ore.

b) Explain why aluminium is so expensive.

SOLVE

Now you have gathered the relevant information it is really important that you think carefully about how you need to use and present that information. Use the prompts below to help you.

Part a) asks you to **describe** how electrolysis is used in industry. You will need to communicate using formulae and equations to show the reactants and products. You could use a labelled diagram in your description.

For part a): Where is aluminium found naturally?

For part a): Why can aluminium not be extracted using carbon like iron can?

For part a): What are the names of the two electrodes used in electrolysis?

Positive electrode:

Negative electrode:

For part a): Add labels to the diagram to show what happens at each electrode in electrolysis.

For part a): Describe or draw how electrolysis is used to extract aluminium. Add as many labels as you can.

For part b): Give two reasons why a lot of energy is used during the extraction of aluminium.

For part b): Why do the anodes need to be replaced regularly?

For part b): Now bring it together: why is aluminium so expensive?

There are additional practice questions with the writing frames for 'Evaluate' at the end of the book.

Session 20 Acids and alkalis

Acids and alkalis are often used in homes in cooking, cleaning and medicines.

Sodium bicarbonate reacts with acids, neutralising them and producing a salt.

Explain why sodium bicarbonate is used in indigestion tablets.

Determine the reactants and products.

THINK

This question needs you to know about the reactions of acids. Can you remember any reactions of acids? What do you know about neutralisation? Scribble down some ideas here.

Now write down the information you will need for this specific question. Not everything you noted down before about neutralisation may be relevant.

Which key idea cards can help? Deal them out next to your work and note extra information here.

Adapt *from Collins*

Don't worry about any blanks. Now have a go at the questions on Adapt© for session 20 to help you think further about this topic.

ACIDS AND ALKALIS

| 1 | 2 | 3 | 4 | 5 | 6 | 7 | 8 | 9 | 10 | 11 | 12 | 13 | 14 |

0–3 = strong acid, 7 = neutral, 12–14 = strong alkali

The **pH** scale is a measure of the acidity or alkalinity of a solution. It can be measured using **universal indicator** or a pH probe. Acids produce hydrogen ions (H^+) in aqueous solutions and have a pH less than 7. Citrus fruit, vinegar and the electrolyte in car batteries are all acidic. Alkaline compounds produce **hydroxide ions** (OH^-) in aqueous solution and have a pH greater than 7. Soap, drain unblocker and oven cleaner are all alkaline.

Acids and alkalis

GENERAL EQUATIONS FOR MAKING SALTS

| acid | + | base | → | a salt | + | water |

| hydrochloric acid | + | copper oxide | → | copper chloride | + | water |

| acid | + | metal | → | a salt | + | hydrogen |

| nitric acid | + | zinc | → | zinc nitrate | + | hydrogen |

| acid | + | metal carbonate | → | a salt | + | water | + | carbon dioxide |

| sulfuric acid | + | magnesium carbonate | → | magnesium sulfate | + | water | + | carbon dioxide |

NEUTRALISATION

Acid	Formula	Alkali	Formula
Hydrochloric acid	HCl	Sodium hydroxide	NaOH
Nitric acid	HNO_3	Potassium hydroxide	KOH
Sulfuric acid	H_2SO_4	Ammonium hydroxide	NH_4OH
Ethanoic acid	CH_3COOH	Calcium hydroxide	$Ca(OH)_2$

When an acid reacts with a base a **neutralisation** reaction takes place and a salt is produced. A **base** is a metal oxide, hydroxide or carbonate. An **alkali** is a base that is soluble in water and releases OH^- ions.

| acid | + | base | → | a salt + water |

| hydrochloric acid | + | sodium hydroxide | → | sodium chloride + water |

The hydrogen ion from the acid reacts with the hydroxide ion from the alkali to make water. Water is neutral and has a pH of 7.

$$H^+ + OH^- → H_2O$$

CONSERVATION OF MASS

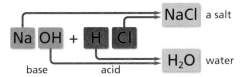

During a chemical reaction, all the atoms are rearranged. This is always true for reactions between acids and alkalis even when a gas is made. Remember, gases have mass. If it looks like the mass has gone down during an experiment, it will be because a gas has escaped into the surroundings.

Count the atoms on each side of the equation. They will always be the same.

sulfuric acid   +   sodium   →   sodium sulfate   +   hydrogen

$$H_2SO_4(aq) \quad + \quad 2Na(s) \quad \rightarrow \quad Na_2SO_4(aq) \quad + \quad H_2(g)$$

nitric acid   +   calcium carbonate   →   calcium nitrate   +   water   +   carbon dioxide

$$2HNO_3(aq) \quad + \quad CaCO_3(s) \quad \rightarrow \quad Ca(NO_3)_2(aq) \quad + \quad H_2O(l) \quad + \quad CO_2(g)$$

During a chemical reaction, atoms are rearranged. All the atoms that were there at the beginning of the reaction still exist in the products. Chemical changes can be communicated using word and formula equations. When **acids** react with **alkalis**, the hydrogen and hydroxide ions rearrange to make water. The other part of the acid and the metal part of the alkali make a **salt**. The name of the salt is a combination of the name of the acid and the metal part of the alkali.

SALTS

Making a salt with excess metal oxide or carbonate

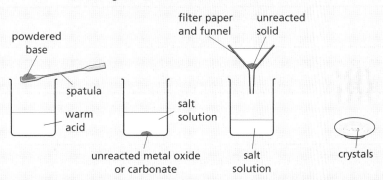

Salt crystals are made when an acid is mixed with excess metal base or carbonate. A base is a metal oxide or alkali. Unreacted solid is removed by filtration and the salt is left to crystallise.

Ions from alkalis and carbonates		Ions from bases and carbonates		Ions from acids	
Sodium	Na^+	Magnesium	Mg^{2+}	Chloride	Cl^-
Potassium	K^+	Zinc	Zn^{2+}	Nitrate	NO^{3-}
Calcium	Ca^{2+}	Copper	Cu^{2+}	Sulfate	SO_4^{2-}

You can work out the name and formula of a salt from the ions present in the acid and base. For example, zinc oxide and sulfuric acid make the salt zinc sulfate. The formula is $ZnSO_4$. The 2+ charge from the Zn^{2+} ion is cancelled out by the 2– charge from the SO_4^{2-} ion.

Session 20   Solving the question

Acids and alkalis are often used in homes in cooking, cleaning and medicines.
Sodium bicarbonate reacts with acids, neutralising them and producing a salt.
Explain why sodium bicarbonate is used in indigestion tablets.
Determine the reactants and products.

SOLVE

Now you have gathered the relevant information it is really important that you think carefully about how you need to use and present that information. Use the prompts below to help you.

This question asks you to use what you know about the reactions of acids to **explain** how excess stomach acid can be treated. You also need to **determine** the reactants and products. This means you need to work out and write down the names of the reactants and products.

What is an acid?

Why does extra stomach acid need treating?

The acid in the stomach can be neutralised by sodium bicarbonate. Write a word equation for the reaction.

The formula for sodium bicarbonate is $NaHCO_3$. Hydrochloric acid is HCl. Write down the formula of the three products. Write how many H atoms there are in a molecule of water. How many C atoms are there in carbon dioxide?

Which product is a gas? What test would you carry out for the gas produced? Describe the test.

The salt produced is called sodium chloride. To practise naming other salts, what is the name of the salt made when hydrochloric acid reacts with magnesium carbonate? Write a word equation for the reaction.

Naming salt practice: What is the name of the salt produced when sulfuric acid reacts with sodium hydroxide? Write a word equation for the reaction.

Now, bring these ideas together to explain why sodium carbonate is used in indigestion tablets and give the reactants and products.

There are additional practice questions with the writing frames for 'Balance', 'Evaluate' and 'Explain' at the end of the book.

Session 21 Rates of reaction

Hydrochloric acid reacts with magnesium to produce magnesium chloride and hydrogen gas.

The graph shows how the volume of hydrogen collected varies with time at two different temperatures of hydrochloric acid.

a) Use collision theory to explain why the reaction is faster when the acid is more concentrated.

b) Suggest another way of increasing the rate of reaction.

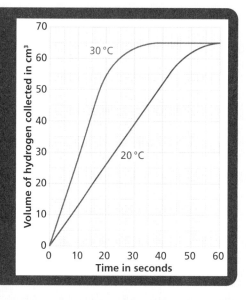

THINK

This question needs you to know about how the rate of reactions can be changed. What can you remember about what happens to the particles of reactants in a chemical reaction? Scribble down some ideas here.

Now write down the information you will need for this specific question. Not everything you noted down before about rates of reaction may be relevant.

Which key idea cards can help? Deal them out next to your work and note extra information here.

Don't worry about any blanks. Now have a go at the questions on Adapt© for session 21 to help you think further about this topic.

COLLISION THEORY

Successful collisions between particles in a reaction

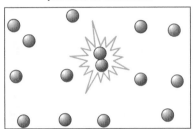

Chemical reactions take place when reactant particles successfully collide with each other and form new products. For a collision to be successful particles must collide in the right direction and with enough energy. The minimum amount of energy needed for a successful collision is called the **activation energy**. The rate of reaction is determined by how many successful collisions there are per second. The more successful collisions, the faster the reaction.

RATES OF REACTION

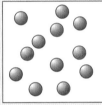 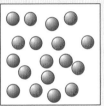

low concentration   high concentration

⚫ reacting particle of substance **A**
⚫ reacting particle of substance **B**

Increasing concentration increases the chances of a successful collision and therefore increases the rate of reaction.

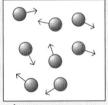 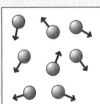

low temperature   high temperature

⚫ reacting particle of substance **A**
⚫ reacting particle of substance **B**

Increasing temperature increases the chances of a successful collision as the particles have more kinetic energy. Increasing temperature therefore increases the rate of reaction.

 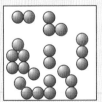

small surface area   large surface area

Increasing surface area increases the chances of particles colliding and so also increases the rate of reaction. A 1 g lump of limestone (calcium carbonate) reacts with hydrochloric acid more slowly than 1 g of powered limestone. This is because in the powdered sample, there are more calcium carbonate particles readily available to collide with the hydrochloric acid.

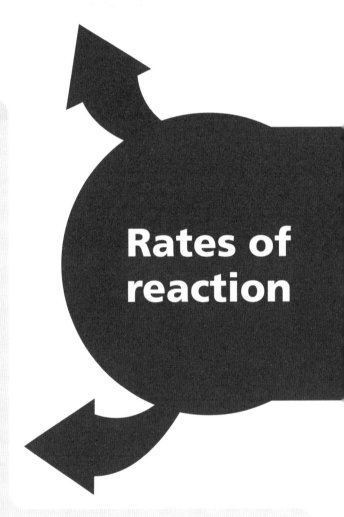

Rates of reaction

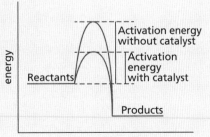

progress of reaction

Catalysts lower the activation energy of a reaction which increases the chances of a successful collision. Catalysts therefore increase the rate of reaction. Catalysts are not used up themselves in the reaction.

TAKING A CLOSER LOOK AT SURFACE AREA

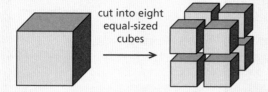

cut into eight
equal-sized
cubes

2 cm × 2 cm × 2 cm cube   1 cm × 1 cm × 1 cm each cube

Smaller particles of the same total mass have a
greater surface area than larger particles.

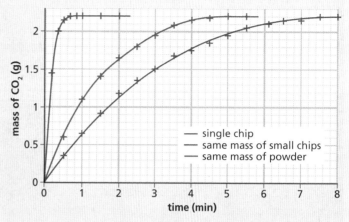

— single chip
— same mass of small chips
— same mass of powder

This graph clearly shows how the rate of reaction increases as
the surface area increases.

A chemical reaction happens when
particles collide with enough energy to
react. The rate of a reaction is a measure
of how fast a product is made or a
reactant is used up. The rate of reaction
gets faster when the concentration of
reactants, the temperature, the surface
area of solid reactants or the pressure of
reacting gases is increased, or a catalyst
is used. A catalyst reduces the amount
of energy the particles need to have to
react when they collide. A catalyst is not
used up in a chemical reaction.

REACTION PROFILES

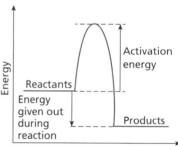

Activation
energy

Reactants

Energy
given out
during
reaction

Products

Reaction profiles show how the energy of
the particles changes during the reaction.
On the profile above the energy level of the
products is lower than the energy level of
the reactants. This means energy was given
out to the surroundings during the reaction.
The reaction is **exothermic**. The profile also
shows the activation energy.

When the energy level of the product
is higher than that of the reactant, the
reaction is **endothermic**. This means
that energy has been taken in from
the surroundings.

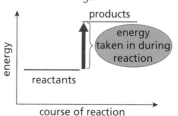

products

energy
taken in during
reaction

reactants

course of reaction

REVERSIBLE REACTIONS

reactants ⇌ products

In some chemical reactions, the products can react to
make the original products again. These reactions are
reversible. When white ammonium chloride is heated,
two colourless gases are formed. When the gases cool,
they reform into white ammonium chloride.

ammonium chloride ⇌ ammonia + hydrogen chloride
(heat / cool)

If the rate of the forward reaction is equal to the rate
of the backward reaction we say that the reaction has
reached **equilibrium**.

Hydrochloric acid reacts with magnesium to produce magnesium chloride and hydrogen gas.

The graph shows how the volume of hydrogen collected varies with time at two different temperatures of hydrochloric acid.

a) Use collision theory to explain why the reaction is faster when the acid is more concentrated.

b) Suggest another way of increasing the rate of reaction.

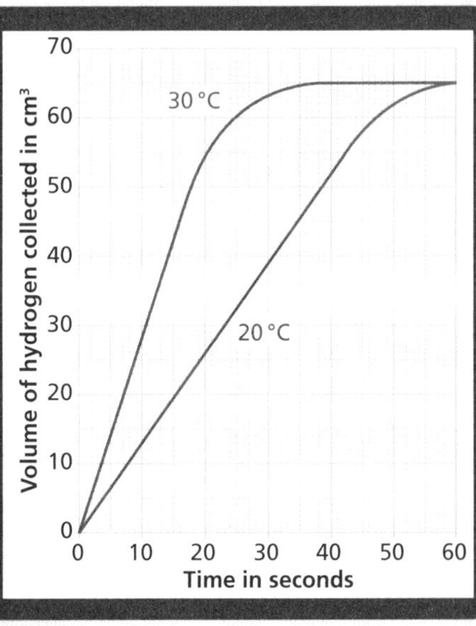

SOLVE

Now you have gathered the relevant information it is really important that you think carefully about how you need to use and present that information. Use the prompts below to help you.

This question asks you to **explain** why a reaction is faster. You need to use what you know about how particles react when there are more particles in the same volume. You then need to **suggest** another way you can change the rate of a reaction. This means you need to use your scientific knowledge to give a suitable answer.

For part a): What happens to particles in a reaction?

For part a): What is the rate of a reaction?

For part a): What happens to the particles in a liquid when you heat it?

For part a): What does this do to the rate of a reaction? Use the graph to support your answer.

For part a): Bring all this together to explain why the rate of a reaction increases when the reactants are heated.

For part b): Now list other ways a reaction can be speeded up.

There are additional practice questions with the writing frames for 'Justify' and 'Suggest' at the end of the book.

Session 22 Chemistry practicals

Acids react with some metals to give a salt and hydrogen. Thermal energy is given out during the reaction.

Design an investigation to show how increasing the concentration of hydrochloric acid affects the temperature change seen as it reacts with zinc metal.

THINK

This question asks you to write about how reactions are changed by increasing the concentration of one of the reactants. You will need to write a method that allows you to measure the temperature change. Without looking at the knowledge organiser, what do you already know about reactions involving acids? Can you remember what happens when the concentration of a reactant is increased? Can you remember why (collision theory)? Can you remember what type of reaction gives out heat to the surroundings? Scribble what you know here, particularly key words and the equipment you would use.

Now write down the information you will need for this specific question. Not everything you noted down before about rates of reaction may be relevant.

Which key idea cards can help? Deal them out next to your work and note extra information here.

Adapt *from Collins*

Don't worry about any blanks. Now have a go at the questions on Adapt© for session 22 to help you think further about this topic.

RISK ASSESSMENT

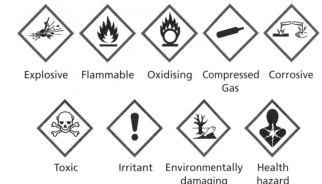

Explosive   Flammable   Oxidising   Compressed   Corrosive
                                        Gas

Toxic   Irritant   Environmentally   Health
                        damaging   hazard

A hazard is anything which could cause harm: for example, hydrochloric acid or water on the floor. A risk is the chance of harm actually happening. That is, how likely it is that you will get acid in your eye, or slip and fall over as you walk through the pool of water on the floor. When planning an experiment, scientists carry out a risk assessment so that they are aware of all the possible hazards and they think about what they can do to minimise the risk. For example, wearing safety glasses and mopping up the water.

For information on chromatography, see Session 17.

PREPARING SOLUBLE SALTS

Stage 1   Mix the reactants and stir. Ensure the copper oxide is in excess

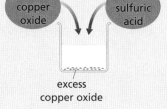

black copper oxide

aqueous sulfuric acid

excess copper oxide

Stage 2   Filter off the excess copper oxide

copper oxide

copper sulfate solution and excess copper oxide

filter funnel

filter paper

filtrate (copper sulfate solution)

Stage 3   Leave the filtrate to evaporate in an evaporating basin or crystallising dish

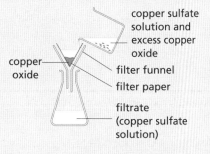

crystallising dish

copper sulfate solution

copper sulfate crystals

The example shows making copper sulfate crystals. When making a soluble salt you always follow the same method. You will need to identify the correct acid and base or carbonate to make the desired salt. To make a purer product you will need to wash, dry and recrystallise the salt.

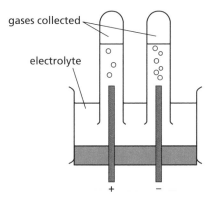

Chemistry practicals

ELECTROLYSIS OF AQUEOUS SOLUTIONS

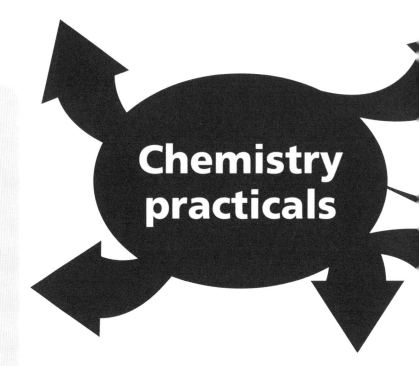

gases collected

electrolyte

+   −

The electrolysis of an aqueous solution is often carried out in an electrolysis cell. Using a small test tube you can collect and test the gases formed at the electrodes. The gas is collected by displacement of the electrolyte from the inverted tubes, so you need to make sure the tubes are full of the liquid at the start. Keep the voltage at around 6 V – no higher.

INVESTIGATING TEMPERATURE CHANGES IN REACTING SOLUTIONS

acid + base → a salt + water + energy

insulated container with lid

reactants

The variables in this investigation include the type of base (remember a base is a metal oxide, metal hydroxide or metal carbonate), the type of acid, the concentration of acid or **alkali** (remember an alkali is a soluble base), the size (or surface area) of the pieces of **metal oxide or carbonate**. During the investigation we can only change one variable, the **independent variable**. All others must be kept the same (**control variables**). In this experiment the **dependent variable** is the temperature change. Use a thermometer or temperature sensor to measure the temperature at the start and the end of the reaction.

INVESTIGATING RATES OF REACTION

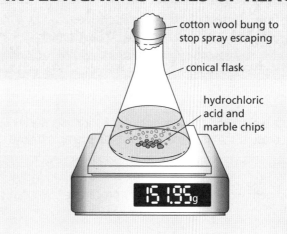

cotton wool bung to stop spray escaping

conical flask

hydrochloric acid and marble chips

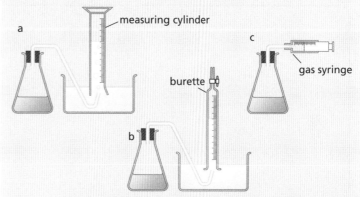

a

measuring cylinder

c

gas syringe

burette

b

The choice of equipment is important when investigating the rate of reaction to get accurate and repeatable results. When hydrochloric acid reacts with marble chips, carbon dioxide gas is produced. We can measure the mass of gas produced over time or the loss in mass of the reactants over time. The mass of gas produced is very small, so a sensitive balance is needed (±0.01 g). There are various ways to collect a volume of gas. Think about the equipment available, how easy it is to set up and the accuracy of the scale.

ANALYSIS AND PURIFICATION OF WATER

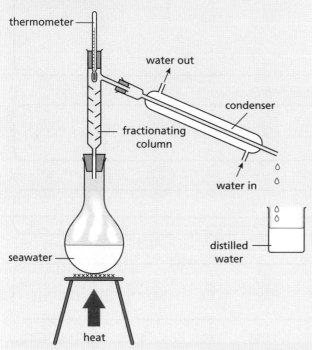

thermometer

water out

condenser

fractionating column

water in

distilled water

seawater

heat

Analytical chemists are like detectives. They have to carry out a series of experiments, process the results and use their knowledge to interpret the results. You need to know which technique to choose. To purify a sample of water, remove any insoluble particles by filtration and then distil most of the filtrate and pour the rest into an evaporating basin and heat it carefully. Any soluble salts will crystallise as the water evaporates.

During distillation, take care when heating, only use a gentle Bunsen flame otherwise dissolved salts in the sample will boil over. Note down the temperature at which the water condenses. It should be 100 °C if it is pure. Test the pH by putting a drop of water onto the universal indicator. It should be 7.

Session 22   Solving the question

Acids react with some metals to give a salt and hydrogen. Thermal energy is given out during the reaction.

Design an investigation to show how changing the concentration of hydrochloric acid affects the temperature change seen as it reacts with zinc metal.

SOLVE

Now you have gathered the relevant information it is really important that you think carefully about how you need to use and present that information. Use the prompts below to help you.

This question asks you to **design** an investigation. This means you need to set out what needs to be done. You need to set out what changing the concentration of the acid does to the temperature of the reaction solution.

What equipment would you use for this procedure? You can show this using a diagram. List the range of concentrations you would use. How will you measure the temperature of the reaction? When will you measure this?

List the steps of the practical procedure.

What are the hazards? What precautions would you take to avoid harm?

What would you expect to happen?

Why would you expect this to happen?

There are additional practice questions with the writing frames for 'Design', 'Draw', 'Plan', 'Show' and 'Sketch' at the end of the book.

Session 23 Forces

Forces can affect the motion of objects.

The badminton shuttlecock shown is falling downwards at constant speed.

a) State the law that gives the relationship between the two forces acting on the shuttlecock when it is moving at a constant speed.

b) Draw arrows on the diagram to show the two forces acting on the shuttlecock.

c) Describe what happens when the shuttlecock hits a stationary badminton racket facing upwards. You can use a diagram to help you explain.

THINK

This question asks you about the effects of forces on an object. Write down the forces that you remember. Think about what happens when forces combine. Remember to think about how forces can be contact forces or non-contact forces. Scribble what you know here.

Now write down the information you will need for this specific question. Not everything you noted down before about forces may be relevant.

Which key idea cards can help? Deal them out next to your work and note extra information here.

Don't worry about any blanks. Now have a go at the questions on Adapt© for session 23 to help you think further about this topic.

FORCES AND THEIR EFFECTS

The thrust of the engines is a push force acting to the right. The **air resistance** between the aircraft and particles of air is a push force acting to the left. The forces are shown as arrows. The two arrows have the same size but act in opposite directions, so they are balanced. This means the aircraft travels at constant speed.

CONTACT AND NON-CONTACT FORCES

The air resistance, in the diagram of the aircraft, is an example of a **contact force**. The air particles hit the surface of the aircraft and oppose the motion of the aircraft. **Friction** and **tension** are also contact forces.

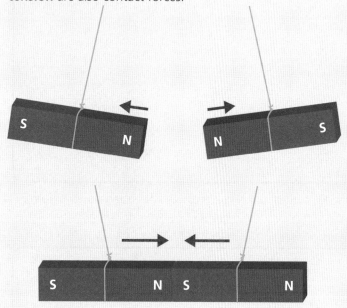

Magnetism is an example of a **non-contact force**, as are gravity and electrostatic forces.

Forces

RESULTANT FORCES AND NEWTON'S SECOND LAW

The diagram shows a driving force of 6000 N to the left. Friction and air resistance combine to produce a force of 2000 N to the right. The **resultant force** is then 4000 N to the left, and the car accelerates to the left. **Newton's Second Law** links the resultant force acting on an object, its mass and its acceleration. The equation is:

resultant force (N) = mass (kg) × acceleration (m/s^2)

If the car has a mass of 2000 kg, then from the equation, 4000 N = 2000 kg × acceleration,

so the acceleration = $\dfrac{4000\,\text{N}}{2000\,\text{kg}}$ = 2 m/s^2.

NEWTON'S FIRST AND THIRD LAWS

drag

friction

forward force from engine

Newton's First Law explains why the car is travelling at constant speed. The forward force from the engine, is equal and opposite to the combined force of the drag and friction. Newton's First Law states that an object will stay still or keep moving at a constant speed unless it is acted on by an external force, so the resultant of these forces on the car is zero.

F_1 F_2

When two cars crash, the green car exerts a force F_2 on the blue car. The blue car exerts an equal and opposite force F_1 on the green car. This is an example of **Newton's Third Law** – when two objects interact, they exert forces on each other that are equal in size and opposite in direction.

A force is a push, pull or twist that causes an object to change speed, change direction or change shape. A force can cause a stationary object to start moving or a moving object to become stationary. When two or more forces act on an object, their effects on the object are added together. If the forces cancel each other out they are said to be balanced, and there is no change in the movement or shape of the object. If the forces are unbalanced, then the movement or shape of the object will change. Newton's First and Second Laws of Motion explain how forces affect a single object. Newton's Third Law of Motion explains that when two objects interact the forces they exert on each other are equal and opposite.

WEIGHT

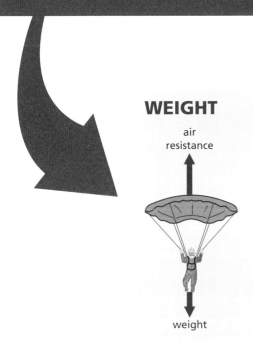

air resistance

weight

The parachutist is falling downwards because the force of gravity is pulling her towards the Earth. This is called her **weight**. The weight of an object, is given by:

weight (N) = mass (kg) × gravitational field strength (N/kg)

The parachute has just opened and has a large air resistance. At this time, the force of air resistance is greater than the weight of the parachutist. Therefore, there is a resultant force acting upwards and the parachutist is slowing down. However, the force of air resistance depends on the speed of movement, so the upwards force reduces as the parachutist slows down. Eventually the weight of the parachutist and the force due to air resistance will be equal in size and opposite in direction, and the parachutist will fall at a constant speed.

Forces can affect the motion of objects.

The badminton shuttlecock shown is falling downwards at constant speed.

a) State the law that gives the relationship between the two forces acting on the shuttlecock when it is moving at a constant speed.

b) Draw arrows on the diagram to show the two forces acting on the shuttlecock.

c) Describe what happens when the shuttlecock hits a stationary badminton racket facing upwards. You can use a diagram to help you explain.

SOLVE

Now you have gathered the relevant information it is really important that you think carefully about how you need to use and present that information. Use the prompts below to help you.

The question first asks you to **draw** arrows to show the forces acting on the shuttlecock. You then need to **describe** what is happening to the shuttlecock. This means you need to remember facts about forces and write in an accurate way.

For part a): Which one of Newton's laws does this show? Decide which one deals with equal and opposite reactions.

For part b): If the shuttlecock is moving at a steady speed then the forces must be balanced. What are the two main forces? Add arrows to the diagram of the shuttlecock to show these forces.

For part c): Try drawing the shuttlecock as it makes contact with the racket. Now add arrows to your diagram to show the two forces. How will you use the size of your arrows to show the size of the forces?

For part c): Describe what happens to the shuttlecock as it makes contact with the racket.

There are additional practice questions with the writing frames for 'Estimate' at the end of the book.

Session 24 Motion

The graph shows a velocity–time graph for a girl riding a bicycle from point A to point E.

a) Describe, in words, the girl's journey from A to E.

b) i) Determine the change in velocity from A to B.

   ii) Calculate the acceleration of the girl from A to B.

c) Calculate the distance that the girl travels between B and C. Use the equation:

   distance (m) = velocity (m/s) × time (s)

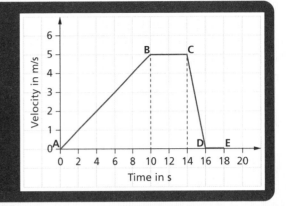

THINK

Graphs can show us a lot about a journey.

What do you remember about velocity–time graphs?

What can you calculate from the graph?

What represents acceleration?

What do flat lines represent?

What is the significance of a line sloping up and of a line sloping down?

Without looking at the knowledge organiser, scribble what you know here about velocity–time graphs and what they show.

Now write down the information you will need for this specific question. Not everything you noted down before about velocity–time graphs may be relevant.

Which key idea cards can help? Deal them out next to your work and note extra information here.

Adapt *from Collins*

Don't worry about any blanks. Now have a go at the questions on Adapt© for session 24 to help you think further about this topic.

SPEED, VELOCITY AND ACCELERATION

Speed is a measure of how fast an object is moving. Speed is a scalar quantity. The equation for speed is:

$$\text{speed (m/s)} = \frac{\text{distance travelled (m)}}{\text{time (s)}}$$

Velocity is the speed of an object in a given direction and it is a vector quantity. **Acceleration** is the rate at which an object changes velocity. Acceleration is a vector quantity. The equation for acceleration is:

$$\text{acceleration (m/s}^2) = \frac{\text{change in velocity (m/s)}}{\text{time taken (s)}}$$

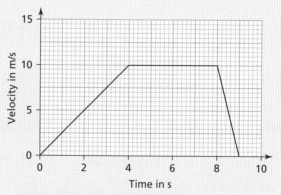

Motion

GRAPHS OF MOTION

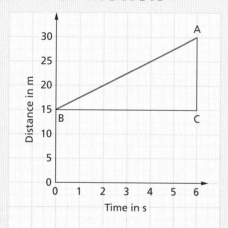

The distance–time graph shows an object moving at constant speed. At first, at point B, the object is 15 m from an observer. It then moves another 15 m further away from the observer in 6 seconds. The **gradient** (or slope) of a distance–time graph is the speed of the object.

$$\text{The gradient of the line} = \frac{AC}{BC}$$
$$= \frac{(30\,\text{m} - 15\,\text{m})}{(6\,\text{s} - 0\,\text{s})}$$
$$= \frac{15\,\text{m}}{6\,\text{s}}$$
$$= 2.5\,\text{m/s}$$

This means that the speed of the object is 2.5 m/s.

Any **horizontal** line on a distance–time graph shows that the object is stationary.

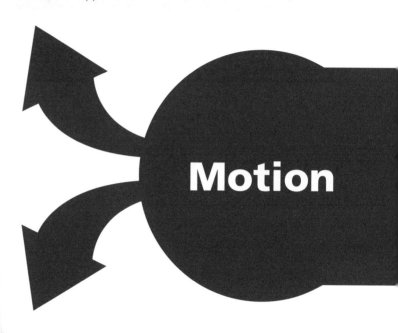

Acceleration can be calculated from the gradient of a velocity–time graph. This example shows the change in velocity of a remote-controlled car during a race. The acceleration between 0 and 4 seconds can be calculated using the gradient of the line.

$$\text{acceleration (m/s}^2) = \frac{\text{change in velocity (m/s)}}{\text{time taken (s)}}$$
$$= \frac{(10 - 0\,\text{m/s})}{4\,\text{s}}$$
$$= 2.5\,\text{m/s}$$

The horizontal part of the graph shows the car moving at a constant velocity of 10 m/s. The downward slope from 8 to 9 seconds shows the car **decelerating** (slowing down).

STOPPING DISTANCES

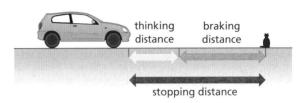

A car's **stopping distance** is the total distance the car travels during the time it takes to stop.

stopping distance (m) = thinking distance (m) + braking distance (m)

Thinking distance depends on the:

- speed of the car
- reaction time of the driver (affected by alcohol/drugs/tiredness/distractions).

Braking distance depends on the:

- speed of the car
- condition of the brakes/tyres
- road surface (dry/wet/icy/loose/firm).

For a given braking force the greater the speed of the vehicle, the greater the stopping distance.

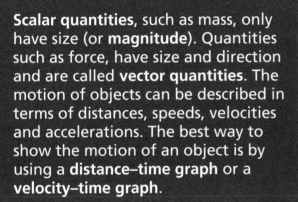

Scalar quantities, such as mass, only have size (or **magnitude**). Quantities such as force, have size and direction and are called **vector quantities**. The motion of objects can be described in terms of distances, speeds, velocities and accelerations. The best way to show the motion of an object is by using a **distance–time graph** or a **velocity–time graph**.

TERMINAL VELOCITY

When a skydiver falls out of a plane, her weight causes the skydiver to accelerate downwards. As the speed of the skydiver increases, so does the air resistance force (the drag). Eventually, the drag force acting upwards, is equal and opposite to the weight acting downwards. The forces are balanced, and the skydiver stops accelerating and travels at a constant speed called the **terminal velocity**.

SCALAR AND VECTOR QUANTITIES

Scalar quantities	Vector quantities
Distance	Displacement
Speed	Velocity
Mass	Weight
Time (interval)	Force
Area, Volume	Acceleration
Density	Lift, Drag, Thrust
Energy	Magnetic field
Work	
Power	
Temperature	

Session 24 Solving the question

> The graph shows a velocity–time graph for a girl riding a bicycle from point A to point E.
>
> a) Describe, in words, the girl's journey from A to E.
>
> b) i) Determine the change in velocity from A to B.
>
>   ii) Calculate the acceleration of the girl from A to B.
>
> c) Calculate the distance that the girl travels between B and C. Use the equation:
>
>   distance (m) = velocity (m/s) × time (s)

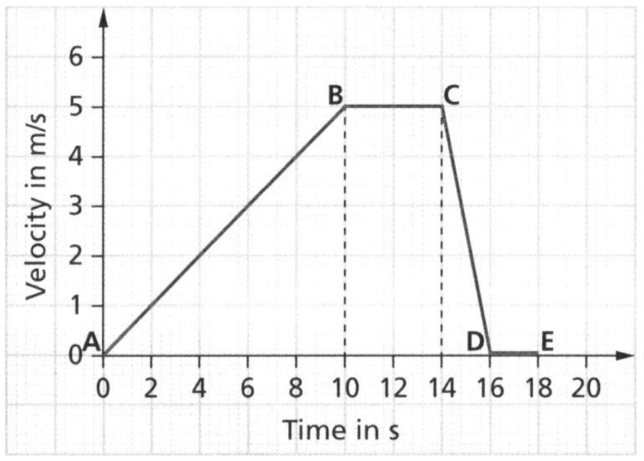

SOLVE

Now you have gathered the relevant information it is really important that you think carefully about how you need to use and present that information. Use the prompts below to help you.

Part a) asks you to **describe** the graph. Your answer must be based on the information given in the graph.

Part b) i) asks you to **determine**, so you must read numbers from the graph.

To **calculate** for parts b) ii) c) you should use numbers from the graph to work out the answer.

To answer part a), break the graph down into steps. In the space below, write what is happening between the labelled points on the graph. The question asks you to use words only so there is no need to write any numbers here.

A to B:

B to C:

C to D:

D to E:

For part b) i) you first need to read values from the graph. Look at the y-axis value after 10 seconds. How much has this value changed from 0 seconds?

For part b) ii), you need to calculate the acceleration from A to B. Put the values for change in velocity and time into the equation: acceleration (m/s²) = change in velocity (m/s) / time taken (s). Remember to write the units.

For part c), what is the constant velocity between B and C? Then calculate the time between B (10 s) and C (14 s). Put these two values into the equation given:

distance (m) = velocity (m/s) × time (s)

What will the units be?

There are additional practice questions with the writing frames for 'Calculate', 'Explain' and 'Plot' at the end of the book.

Session 25 Waves

The diagram shows a transverse water wave.
The frequency of the wave is 3 Hz and the wavelength of the wave is 2 m.

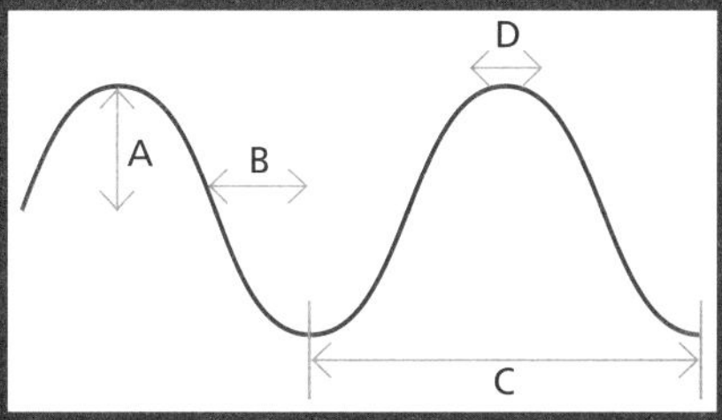

a) Identify the label corresponding to the:

   i) amplitude

   ii) wavelength.

b) Calculate the speed of the wave using the wave equation.

c) The table below shows the parts of the electromagnetic spectrum. Some of the parts are missing.

radio waves		infrared		ultraviolet		gamma rays

   i) Write the missing words in the table.

   ii) Draw an arrow under the table to show in which direction wavelength increases.

   iii) Explain why ultraviolet rays are dangerous to humans.

THINK

Waves transfer energy. Can you remember the two types of wave? Waves have an amplitude, frequency and wavelength. Can you remember how to use these values to calculate a velocity (speed)? Frequency and wavelength vary greatly across the electromagnetic spectrum. Without looking at the knowledge organiser, what can you remember about the electromagnetic spectrum? Scribble here what you know about waves, their properties and the electromagnetic spectrum. Include any equations you can remember.

Now write down the information including all equations you will need for this specific question. Not everything you noted down before about waves may be relevant.

Which key idea cards can help? Deal them out next to your work and note extra information here.

Adapt *from Collins*

Don't worry about any blanks. Now have a go at the questions on Adapt© for session 25 to help you think further about this topic.

105

WAVES

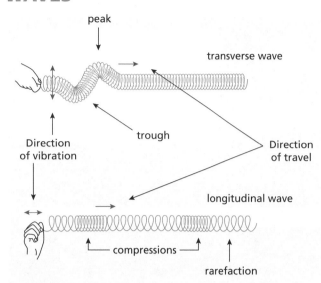

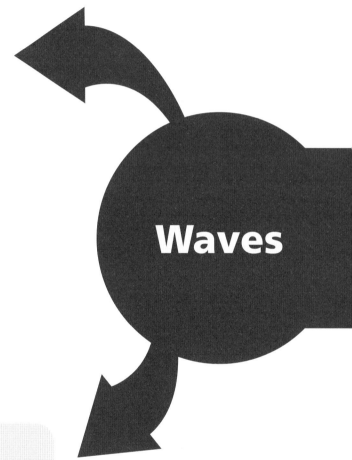

Waves

The diagram shows a transverse wave and a longitudinal wave being generated on a slinky. For the transverse wave, the coils move at right angles to the direction that the wave is travelling. This causes peaks and troughs to form. For the longitudinal wave, the coils move backwards and forwards, forming **compressions** and **rarefactions** in the same direction as the direction of travel. All waves transfer energy.

PROPERTIES OF WAVES

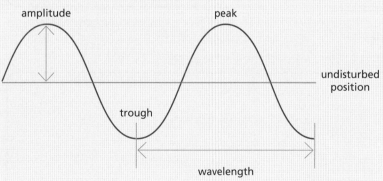

The diagram shows two key properties of a wave. The **wavelength** (in m) is the distance from one point on a wave to the equivalent point on the next wave. The **amplitude** (in m) is the maximum displacement of a point on a wave from its undisturbed position. The **frequency** (in hertz, Hz) of a wave is the number of complete waves passing a point in 1 second. Frequency and wavelength are connected to the wave speed (in m/s) by the wave equation:

wave speed (m/s) = frequency (Hz) × wavelength (m)

THE ELECTROMAGNETIC SPECTRUM

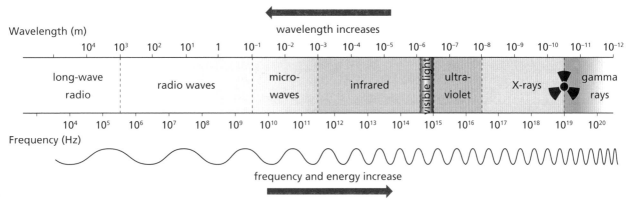

Wavelength (m)

wavelength increases

| 10^4 | 10^3 | 10^2 | 10^1 | 1 | 10^{-1} | 10^{-2} | 10^{-3} | 10^{-4} | 10^{-5} | 10^{-6} | 10^{-7} | 10^{-8} | 10^{-9} | 10^{-10} | 10^{-11} | 10^{-12} |

long-wave radio | radio waves | micro-waves | infrared | visible light | ultra-violet | X-rays | gamma rays

| 10^4 | 10^5 | 10^6 | 10^7 | 10^8 | 10^9 | 10^{10} | 10^{11} | 10^{12} | 10^{13} | 10^{14} | 10^{15} | 10^{16} | 10^{17} | 10^{18} | 10^{19} | 10^{20} |

Frequency (Hz)

frequency and energy increase

The **electromagnetic spectrum** is a family of transverse waves. The waves all travel at the speed of light through a vacuum, and transfer energy from a source to an absorber. They are arranged into seven groups by order of wavelength, frequency and energy. In decreasing order of wavelength (and increasing frequency/energy), they are: **radio waves**, **microwaves**, **infrared**, **visible light**, **ultraviolet**, **X-rays** and **gamma rays**.

Waves transfer energy. **Longitudinal waves** (e.g. sound waves) cause the particles of the substance the wave is travelling through to move backwards and forwards. **Transverse waves** (e.g. water waves) make the particles move at a right angle to the direction the wave is travelling. Different parts of waves can be measured to understand the properties of waves.

PROPERTIES AND USES OF THE ELECTROMAGNETIC SPECTRUM

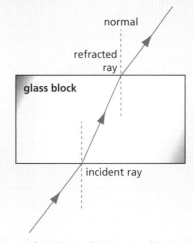

The diagram shows the **refraction** of a beam of light through a glass block. All waves undergo refraction. Refraction allows infrared and visible light to be used in fibre optics. Radio waves, microwaves, infrared and visible light are all used in communications; and microwaves and infrared are both used for cooking. Ultraviolet, X-rays and gamma rays have the highest energies and can be hazardous to the human body, harming or killing cells. X-rays and gamma rays can be used for medical treatments and imaging. The effects of these waves depend on the size of the dose received by the body.

Session 25 Solving the question

The diagram shows a transverse water wave.
The frequency of the wave is 3 Hz and the wavelength of
the wave is 2 m.

a) Identify the label corresponding to the:
  i) amplitude
  ii) wavelength.
b) Calculate the speed of the wave using the wave equation.
c) The table below shows the parts of the electromagnetic spectrum. Some of the parts
   are missing.

radio waves		infrared		ultraviolet		gamma rays

  i)   Write the missing words in the table.
  ii) Draw an arrow under the table to show in which direction wavelength increases.
  iii) Explain why ultraviolet rays are dangerous to humans.

SOLVE

Now you have gathered the relevant information it is really important that you think carefully about how you need
to use and present that information. Use the prompts below to help you.

Part a) asks you to **identify** key features on the diagram. This means you need to select the correct letter labels
from the diagram.

In part b) you need to **calculate** the speed of the wave. You should use numbers given in the question to work out
the answer.

You then need to recall information about the structure of the electromagnetic spectrum. For **explain** questions
like part c) iii), you should make something clear, or state the reasons for something happening.

For part a): What are the characteristics of a transverse (e.g. water) wave? Where do we measure amplitude and
wavelength from?

> i) amplitude =
> ii) wavelength =

For part b): Gather all the numerical information from the question that you need to use the wave equation:

wave speed (m/s) = frequency (Hz) × wavelength (m)

Then substitute the numbers into the equation and use a calculator to work out the answer. Repeat this on your
calculator to check that you have entered the numbers correctly. What are the units of wave speed in this question?

For part c) i), ii): List all of the parts of the electromagnetic spectrum in the correct order. Try to remember which
have the longest wavelength, and so which way the arrow should point.

For part c) iii): You need to **explain** in this question – this means you need to state the reasons for something.
First, state the property of ultraviolet rays that makes them dangerous. Next, give the reasons why this makes
them dangerous.

There are additional practice questions with the writing frames for 'Estimate' and 'Show' at the end of the book.

Session 26 Energy

A man, with a mass of 75 kg, climbs a flight of steps 1.5 m high. At the top of the steps he stops, before jumping off the top step and landing back down on the ground.

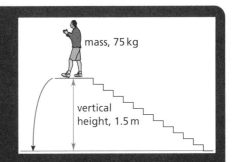

mass, 75 kg

vertical height, 1.5 m

a) Describe the changes in energy as he climbs the steps, stops, and then jumps down.

b) Calculate the gravitational potential energy of the man at the top of the steps.
   The gravitational field strength is 10 N/kg.

c) The velocity of the man, just before he hits the floor is 5.2 m/s.
   Calculate the kinetic energy of the man just before he hits the floor.

d) Use your answers to parts b) and c) to calculate the energy wasted as sound energy and thermal energy as the man falls.

THINK

This question needs you to think about energy stores and energy transfers.

Energy stores come in many forms, including kinetic energy, gravitational potential energy, internal energy (including thermal energy and chemical energy), elastic potential energy and electrical energy. Energy can be transferred from one store into other stores via heating, moving, mechanical forces, electricity or by radiation. The law of conservation of energy states that energy can never be created or destroyed but it can be transferred from one store into other stores.

Without looking at the knowledge organiser, what do you already know about energy transfers? Can you remember any of the equations for calculating amounts of energy? Scribble down what you know about energy transfers, including any equations and units.

Now write down the information, including all equations, you will need for this specific question. Not everything you noted down before about energy changes may be relevant.

Which key idea cards can help? Deal them out next to your work and note extra information here.

Don't worry about any blanks. Now have a go at the questions on Adapt© for session 26 to help you think further about this topic.

ENERGY STORES

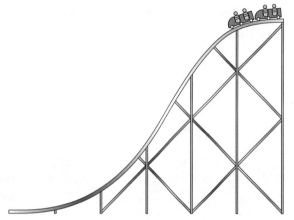

Energy can be stored in different forms, including kinetic energy and gravitational potential energy. The kinetic energy of an object depends on:

- the mass of the object
- and its speed.

The equation for kinetic energy is:

$$\text{kinetic energy (J)} = 0.5 \times \text{mass (kg)} \times \text{speed (m/s)}^2$$

$$E_k = 0.5 \times m\,v^2$$

The gravitational potential energy of an object depends on:

- the mass of the object
- its height above the ground
- and the **gravitational field strength**.

The equation for gravitational potential energy is:

$$\text{gravitational potential energy (J)} = \text{mass (kg)} \times \text{gravitational field strength (N/kg)} \times \text{height (m)}$$

$$E_p = m\,g\,h$$

Energy

ENERGY TRANSFERS

Energy can be transferred from one store to other stores in different ways, such as:

- by heating
- electrically
- by radiation
- mechanically.

The gravitational potential energy of a falling object can be mechanically converted into kinetic energy. If all the gravitational potential energy is converted into kinetic energy, then:

$$\tfrac{1}{2} \times \text{mass} \times \text{speed}^2$$
$$= \text{mass} \times \text{gravitational field strength} \times \text{height}$$

POWER AND EFFICIENCY

Power is the rate of transfer of energy. The equation for power is:

$$\text{power (W)} = \frac{\text{energy transferred (J)}}{\text{time (s)}}$$

When energy transfers from one store into other stores, some energy is always wasted, usually through heating. This energy is not lost but is transferred to a less useful energy store. The **efficiency** of an energy transfer is given by the equation:

$$\text{efficiency} = \frac{\text{useful output energy transfer}}{\text{total input energy transfer}}$$

or, in terms of power:

$$\text{efficiency} = \frac{\text{useful power output}}{\text{total power input}}$$

Energy stores are needed for actions. Energy is transferred from one store to another during an action. A force causes the transfer of energy from one store to another. For example, when a ball is dropped the gravitational potential energy stored in the ball is transferred to kinetic energy. The rate of transfer of energy is called **power**. More powerful devices can transfer more energy per second.

ENERGY RESOURCES

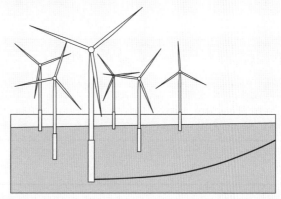

Energy resources are generally classified into **non-renewable resources** (coal, oil, gas and nuclear) and **renewable resources** (biofuels, hydroelectricity, geothermal, tidal, solar, wind and wave energies). Energy resources are used for heating, transport and electricity generation.

When choosing an energy resource for a purpose, the following should be considered:

- the environmental issues associated with the resource
- the reliability of the resource
- any economic considerations.

Session 26 Solving the question

A man, with a mass of 75 kg, climbs a flight of steps 1.5 m high. At the top of the steps he stops, before jumping off the top step and landing back down on the ground.

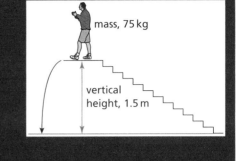

mass, 75 kg

vertical height, 1.5 m

a) Describe the changes in energy as he climbs the steps, stops, and then jumps down.

b) Calculate the gravitational potential energy of the man at the top of the steps.
The gravitational field strength is 10 N/kg.

c) The velocity of the man, just before he hits the floor is 5.2 m/s.
Calculate the kinetic energy of the man just before he hits the floor.

d) Use your answers to parts b) and c) to calculate the energy wasted as sound energy and thermal energy as the man falls.

SOLVE

Now you have gathered the relevant information it is really important that you think carefully about how you need to use and present that information. Use the prompts below to help you.

Part a) asks you to **describe** the energy changes as the man walks up the steps and then jumps down. This means that you will need to remember some key facts and processes about energy stores and transfers. In parts b), c) and d) you need to **calculate** the gravitational potential energy, kinetic energy and energy wasted. For these questions you should use numbers given in the question and/or the diagram to work out the answers.

For part a), you need to describe how the energy transfers from kinetic energy into gravitational potential energy and back again. How is energy wasted?

For part b), you need to recall the equation for gravitational potential energy:

gravitational potential energy = mass × gravitational field strength × height

Underline all the values you have been given in the question and/or on the diagram. Now insert them into the equation. Can you remember the units? Use a calculator to work out the answer. Remember to check your answer by repeating the calculation on your calculator.

For part c), you need to recall the equation for kinetic energy:

kinetic energy = 0.5 × mass × velocity^2

Underline all the values you have been given in the question and/or on the diagram. Now insert them into the equation. Can you remember the units? Use a calculator to work out the answer. Remember to check this by repeating the calculation on your calculator.

For part d), remember that the law of conservation of energy states that energy does not just disappear. Therefore, the gravitational potential energy at the top of the steps is transferred into kinetic energy plus wasted energy.

energy wasted = gravitational potential energy − kinetic energy

Use the values from parts b) and c) in your calculation. Remember to write the units. Now use a calculator to work out the answer. Remember to check this by repeating the calculation on your calculator.

There are additional practice questions with the writing frames for 'Compare' and 'Describe' at the end of the book.

Session 27 Electricity

A student performs an experiment to measure the potential difference across, and current through, a fixed resistor. She collects the following results:

Potential difference in V	0	2	4	6	8	10
Current in A	0.0	0.1	0.2	0.5	0.4	0.5

a) Plot a graph of these results. Plot the potential difference on the *x*-axis and current on the *y*-axis.

b) One of the results has been recorded incorrectly and is an anomaly. Identify the anomaly and draw a circle around this result on your graph.

c) Draw a line of best fit through the valid data points.

d) Suggest a correct value for the anomalous result.

e) The equation for resistance is:

$$\text{resistance} = \frac{\text{potential difference}}{\text{current}}$$

Calculate the resistance of the fixed resistor.

Use the result for a potential difference of 8 V and the equation above.

THINK

This question is about calculating the resistance of a fixed resistor from data supplied in a table. You need to use the data to plot a graph. What can you remember about resistance and how is it related to the other quantities given in the table? How do you plot a good graph? Scribble down some ideas here.

Now write down the information that you will need for this specific question. Not everything you noted down about resistance and plotting graphs may be relevant.

Which key idea cards can help? Deal them out next to your work and note extra information here.

Adapt *from* Collins

Don't worry about any blanks. Now have a go at the questions on Adapt© for session 27 to help you think further about this topic.

ELECTRICAL CIRCUITS

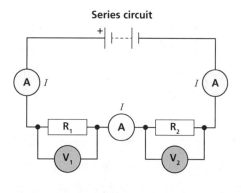

Series circuit

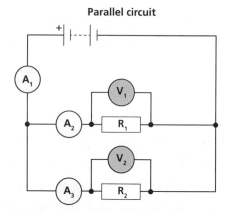

Parallel circuit

In the series circuit, components, like **resistors**, are connected in a closed loop. The current, I, is the same through each component. The total **potential difference** (voltage) of the power supply, V, is shared between the components, so $V = V_1 + V_2$. In the parallel circuit, components are connected across each other, using junctions. The potential difference across each component is the same, so $V_1 = V_2$. The total current flowing through the power supply, I_1, is the sum of the currents through each component, or $I_1 = I_2 + I_3$.

ELECTRICAL POWER

The power of an electrical device is the rate at which it converts energy supplied in the form of electricity into other more useful forms of energy, such as light, heat or kinetic energy. Power is measured in watts (W).

Power is given by the equation:

power (W) = potential difference (V) × current (A)

Power can also be worked out using current and resistance:

power = (current)² (A)² × resistance (Ω)

Electricity

RESISTANCE

The resistance of a component is its opposition to the flow of current through it. Resistance is measured in ohms (Ω). Resistance can be calculated using the equation:

potential difference (V)   =   current (A)   ×   resistance (Ω)

$$V \quad = \quad I \quad \times \quad R$$

In the series circuit, the total resistance, R, of the circuit is the sum of resistances, R_1 and R_2, or $R = R_1 + R_2$.

In the parallel circuit, the total resistance is less than the resistance of the lowest individual resistor.

RESISTANCE OF A WIRE

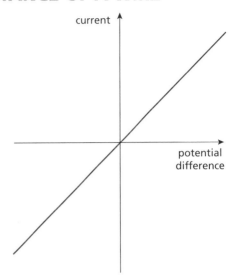

An *I–V* (current–potential difference) graph for a length of wire is shown. The wire obeys the equation $V = IR$, with a constant value of Resistance, R. A wire with a lower resistance would have a steeper gradient (slope). The resistance of a wire depends on the length of the wire: a longer wire has a higher resistance than a shorter similar wire. If the length of the wire is double, then its resistance doubles.

An electrical circuit is a closed loop of conducting material with at least one **cell** and a component (e.g. a lamp). The cell is the source of electrical energy. Electric **current** is the rate of flow of **charge** through the circuit. As current flows, it is opposed by the structure of the circuit causing **resistance**. In a **series circuit**, the current is the same all the way round the circuit. A **parallel circuit** has different loops for the current to flow through. The total current of all the loops in a parallel circuit is equal to the current flowing through the cell.

MAINS ELECTRICITY

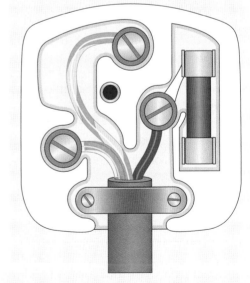

Mains electricity is supplied as an **alternating current** (ac), at a frequency of 50 Hz and a potential difference of 230 V. Mains cables have three coloured wires:

- brown – the live wire (which carries the potential difference).
- blue – the neutral wire (which completes the circuit).
- green/yellow stripes – the earth wire (a safety wire, that stops appliances from becoming live when there is a fault).

Session 27 Solving the question

A student performs an experiment to measure the potential difference across, and current through, a fixed resistor. She collects the following results:

Potential difference in V	0	2	4	6	8	10
Current in A		0.0	0.1	0.2	0.5	0.4	0.5

a) Plot a graph of these results. Plot the potential difference on the *x*-axis and current on the *y*-axis.

b) One of the results has been recorded incorrectly and is an anomaly. Identify the anomaly and draw a circle around this result on your graph.

c) Draw a line of best fit through the valid data points.

d) Suggest a correct value for the anomalous result.

e) The equation for resistance is:

$$\text{resistance} = \frac{\text{potential difference}}{\text{current}}$$

Calculate the resistance of the fixed resistor.
Use the result for a potential difference of 8 V and the equation above.

SOLVE

Now you have gathered the relevant information it is really important that you think carefully about how you need to use and present that information. Use the prompts below to help you.

This question first asks you to **plot** which means you have to mark the points on the graph using the data given. You then need to **identify** the anomaly and **suggest** a value that fits the pattern. The final part asks you to **calculate** the resistance using the equation given and the numbers read from the graph.

For part a) you need to **plot** a graph. First, look at the data given. You are told which row needs to be plotted on the *x*-axis and which one on the *y*-axis. Does the graph start at (0, 0)? What are the maximum *x*-axis and *y*-axis values? Draw a scale that fits both using a pencil and a ruler (in case you make a mistake). Plot the data points accurately using the scales. Remember to label your axes using the table row headings.

For part b), can you see an **anomaly** (a data point that does not fit the pattern)? Circle it on your graph.

For part c), think about lines of best fit. Is this one a straight line, or a curve? Does it go through (0, 0)? Do you include the anomaly?

For part d), now use the line of best fit that you have drawn to identify the true value of the current at 6 V.

For part e), read the value of the current off the graph corresponding to a potential difference of 8 V. Write down the data values that you have read off the graph, and the equation given. Insert the values into the correct place in the equation and use a calculator to calculate your answer. Repeat the calculation using your calculator to double check your answer. Write down your final answer and write down the unit of resistance. You can use the symbol or the word.

There are additional practice questions with the writing frames for 'Define' and 'Draw' at the end of the book.

Session 28 Electromagnetism

A child's toy magnet set, containing three solid metal rods, has become mixed up.
- One rod is a permanent magnet.
- The second rod is made from iron.
- The third rod is made from aluminium.

All the rods look the same.

Explain how a permanent bar magnet could be used to determine the nature of each rod.

THINK

This question is about using your knowledge of magnets and magnetic fields to identify the three objects. Without looking at the knowledge organiser, what do you already know about magnets and magnetic fields? What can you remember about how magnetic poles interact and about induced magnetism? Scribble down some notes here.

Now write down the information you will need for this specific question. Not everything you noted down before about magnets and magnetic fields may be relevant.

Which key idea cards can help? Deal them out next to your work and note extra information here.

Adapt *from Collins*

Don't worry about any blanks. Now have a go at the questions on Adapt© for session 28 to help you think further about this topic.

MAGNETIC POLES

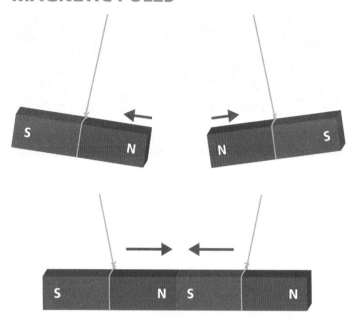

Permanent magnets produce their own magnetic field, and they have a north pole and a south pole. When they are brought together, magnetic **poles** exert a force on each other. Like poles **repel** each other and unlike poles **attract** each other. A permanent magnet can make objects that are made of the materials iron or steel, magnetic – this is called induced magnetism. Induced magnetism always causes a force of attraction. If the permanent magnet is moved away from the **induced magnet**, the induced magnet quickly loses its magnetism.

Electromagnetism

MAGNETIC FIELDS AROUND MAGNETS

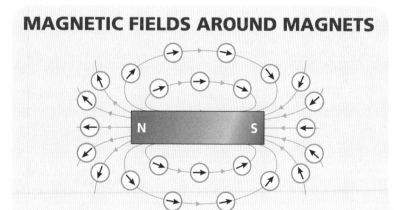

A magnetic field is a region around a magnet where another magnet (or an iron/steel object) experiences a force. **Magnetic field lines** can be identified using a plotting compass. Magnetic field lines can be drawn using arrowed lines that point away from north poles and towards south poles. Magnetic fields are strongest where the magnetic field lines are closest together, i.e. at the poles. Magnetic fields get weaker as the distance away from the magnet increases. The Earth has a permanent magnetic field. A magnetic compass has a magnetised needle, and one end of the needle always points towards the Earth's north pole.

ELECTROMAGNETS

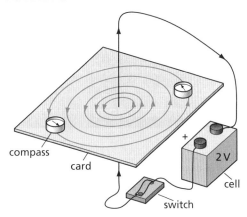

When an electric current flows through a metal wire, a magnetic field is produced around the wire. The magnetic field lines form **concentric circles** around the wire. Coiling the wire produces an electromagnet. The strength of the magnetic field around an electromagnet depends on the size of the electric current, the number of turns of the coil of wire and the distance away from the electromagnet. A **solenoid** is a long, coiled electromagnet. The magnetic field inside the solenoid is **uniform** and can be increased in strength by adding an iron core.

Magnetism is a non-contact force between two poles of a **magnet**. The **magnetic fields** around magnets can be drawn using magnetic field lines. The magnetic field is strongest at the poles of a magnet. **Electromagnets** are formed when an electric current flows through a conductor such as a metal wire. The shape of the magnetic field around an electromagnet depends on the shape of the electromagnet.

MAGNETIC FIELDS AROUND ELECTROMAGNETS

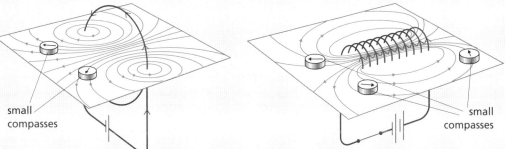

The diagram shows the shape of the magnetic fields around a single coil (on the left) and a solenoid (on the right). The magnetic field of the solenoid is like the magnetic field around a permanent magnet. One end of the solenoid acts like a north pole and the other end like a south pole. Electromagnets have advantages over permanent magnets because the magnetic fields can be turned on and off. Also, electromagnets can be varied in strength by varying the electric current.

Session 28 Solving the question

A child's toy magnet set, containing three solid metal rods, has become mixed up.
- One rod is a permanent magnet.
- The second rod is made from iron.
- The third rod is made from aluminium.

All the rods look the same.

Explain how a permanent bar magnet could be used to determine the nature of each rod.

SOLVE

Now you have gathered the relevant information it is really important that you think carefully about how you need to use and present that information. Use the prompts below to help you.

This question asks you to **explain**. This means that you need to state, with reasons, the steps that you should take to identify each rod.

What are the rules for the interaction of magnetic poles? How could these be used to identify the permanent magnet rod?

What is induced magnetism? What materials can be made into an induced magnet? Does it matter which pole of the permanent magnet is used to induce magnetism? How could this be used to identify the iron rod?

What happens when a permanent magnet is brought up close to a material that cannot be magnetised? How could this be used to identify the aluminium rod?

Now bring together all this information to explain how to determine the nature of each rod.

There are additional practice questions with the writing frames for 'Predict' at the end of the book.

Session 29 Atoms and radioactivity

The table below shows how the activity of a radioactive sample of lead-217 (atomic number 82) changes over time.

Time in s	0	10	20	30	40	50
Activity in Bq	120	85	60	42	30	21

a) Plot a graph of activity against time.

b) Draw a line of best fit through the points.

c) Use information from the graph to determine the half-life of lead-217.

d) Lead-217 decays by beta particle emission into an isotope of bismuth, Bi. Balance the nuclear decay equation for this decay:

$$^{217}_{82}\text{Pb} \rightarrow {}^{0}_{-1}\text{e} + {}^{...}_{...}\text{Bi}$$

THINK

This question is about using your knowledge of radioactive decay to determine the half-life of lead-217. Without looking at the knowledge organiser, what do you already know about radioactive decay? What can you remember about how to determine the half-life of a radioactive substance? What can you remember about the equations for beta particle decay? Scribble down some notes here.

Now write down the information you will need for this specific question. Not everything you noted down before about radioactive decay may be relevant.

Which key idea cards can help? Deal them out next to your work and note extra information here.

Adapt *from Collins*

Don't worry about any blanks. Now have a go at the questions on Adapt© for session 29 to help you think further about this topic.

THE ATOMIC MODEL

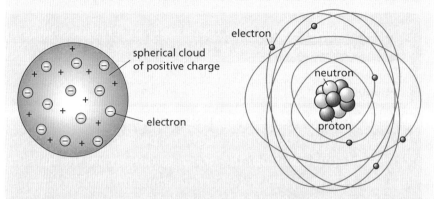

ISOTOPES

Isotopes of elements are atoms with the same number of protons but different numbers of neutrons. The nuclei of different isotopes can be represented by the symbols $^A_Z X$, where A is the mass number, which is equal to the proton number + neutron number; Z is the atomic number, or the proton number; and X is the chemical symbol. The diagram shows the isotope uranium-238 – it has 92 protons and 146 neutrons. An isotope of lithium can be represented as $^7_3 Li$ – it has 3 protons (the atomic number); and 4 neutrons in its nucleus, so the mass number is 7.

Our model of the atom has changed over time. The diagram on the left shows the 'plum pudding' model from 1904. Alpha particle scattering experiments in 1911 showed that this model was incorrect, and that atoms have a tiny nucleus surrounded by electrons. This model was adapted further in 1913 to include the electrons orbiting the nucleus, as shown in the right-hand diagram. Protons were discovered in 1917 and neutrons in 1932. New experimental evidence often leads to a scientific model being changed or replaced.

Atoms and radioactivity

NUCLEAR RADIATION

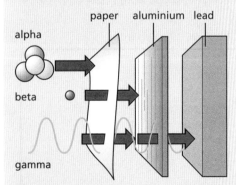

During radioactive decay, unstable nuclei can emit radiation and become stable. The rate of radioactive decay is called activity (measured in **becquerel**, Bq), and can be measured using a Geiger–Muller tube. The diagram shows three of the four types of nuclear radiation. **Alpha particles** are helium nuclei and are absorbed by a thin piece of paper. **Beta particles** are high energy electrons emitted from the nucleus. Beta particles are absorbed by a few millimetres (mm) of aluminium. Gamma rays are electromagnetic radiation emitted by the nucleus and are absorbed by a few centimetres (cm) of lead. Neutrons can also be emitted by nuclei.

NUCLEAR EQUATIONS

The diagram on the left represents an example of alpha particle radioactive decay. In this case, a nucleus of uranium-238 decays via the emission of an alpha particle, $^4_2$He, giving a daughter nucleus of thorium-234. During beta particle radioactive decay (diagram on the right), a nucleus of carbon-15 emits a beta particle, $^0_{-1}$e, leaving a daughter nucleus of nitrogen-15. In general:

$$\alpha \text{ decay } ^A_Z X \rightarrow {}^4_2 He + {}^{A-4}_{Z-2} Y$$
$$\beta \text{ decay } ^A_Z X \rightarrow {}^0_{-1} e + {}^A_{Z+1} Y$$

Atoms are the tiny, basic building blocks of **matter**. The nuclear model of the atom is a small nucleus containing protons and neutrons, surrounded by orbiting electrons. **Isotopes** of elements have the same number of protons but different numbers of neutrons. Some atoms are unstable and **emit** nuclear radiation in a random process called **radioactive decay**. Nuclear **radiation** is hazardous to living cells and precautions need to be taken when handling **radioactive** substances.

RADIOACTIVE DECAY

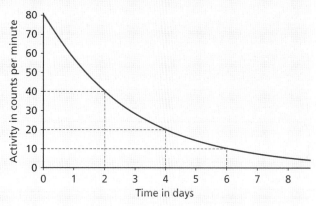

Although radioactive decay is random, the pattern is very predictable. The **half-life** of a radioactive substance is the time taken for the number of radioactive nuclei (or the activity) to fall to half its initial value. The half-life of an isotope is always constant, but different isotopes can have different half-lives varying from tiny fractions of a second to many trillions of years. In the diagram, the half-life is 2 days. Radioactive contamination of materials occurs when radioactive substances combine with other materials. Irradiation is the process of exposing an object to nuclear radiation.

Session 29 Solving the question

The table below shows how the activity of a radioactive sample of lead-217 (atomic number 82) changes over time.

Time in s	0	10	20	30	40	50
Activity in Bq	120	85	60	42	30	21

a) Plot a graph of activity against time.
b) Draw a line of best fit through the points.
c) Use information from the graph to determine the half-life of lead-217.
d) Lead-217 decays by beta particle emission into an isotope of bismuth, Bi. Balance the nuclear decay equation for this decay:

$$^{217}_{82}Pb \rightarrow ^{0}_{-1}e + ^{...}_{...}Bi$$

SOLVE

Now you have gathered the relevant information it is really important that you think carefully about how you need to use and present that information. Use the prompts below to help you.

In this question you first need to **plot** a graph of the data given. In this case you need to design and plot the axes and the data. You are then asked to **draw** a line of best fit. This means that you need to add a line to your graph. You then need to **determine** the half-life of lead-217, using information from the graph. Finally, you need to **balance** the nuclear decay equation for the decay of lead-217 via beta particle emission.

For part a): You will first need to determine the axes for the decay graph. Which quantity goes on which axis? What scale do you need – what are the lowest and the highest values that you need to plot on each axis? Do you need to start the graph axes at (0, 0)? Make sure that you plot the points accurately *using a sharp pencil*. Double-check them.

For part b): Think about the general shape plotted out by the points (is it a curve or a straight line)? Remember, a line of best fit does not need to go through all the points, but it does need to be a smooth line, not join-the-dots.

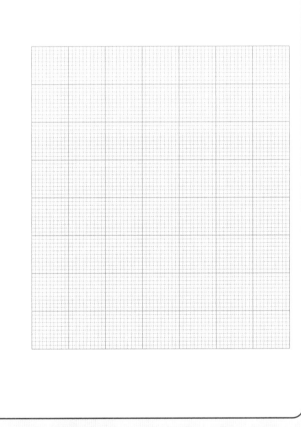

For part c): The half-life is the time taken for the initial activity to fall to half its value. What is the initial activity? What is half of this initial value? Draw a straight line across the graph from this value, so it touches the line of best fit. Then draw a line down to the time axis and read off the half-life. What are the units?

For part d): You need to balance the nuclear decay equation. The top, mass numbers need to be the same on each side of the equation. What is the mass number of the bismuth isotope? The bottom, atomic numbers, also need to balance on either side of the equation. Remember the value for a beta particle is −1, so the atomic number of bismuth − 1 = atomic number of lead. What is the atomic number of the bismuth isotope?

There are additional practice questions with the writing frames for 'Compare', 'Define', 'Determine' and 'Justify' at the end of the book.

Session 30 Physics practicals

Design an investigation that shows the effect of increasing the mass of an object hung from a spring.

THINK

This question asks you to write about springs and what happens when the mass is changed. You will need to write a method that includes different masses and a way of measuring the extension of the spring. Without looking at the knowledge organiser, what can you remember about springs? Can you remember what you would use to measure length? Which unit would you measure the length in? Scribble what you know here, particularly key words, the equation for the calculation and the equipment you would use.

Now write down the information you will need for this specific question. Not everything you noted down before about investigating springs may be relevant.

Which key idea cards can help? Deal them out next to your work and note extra information here.

Adapt *from* Collins

Don't worry about any blanks. Now have a go at the questions on Adapt© for session 30 to help you think further about this topic.

INVESTIGATING SPECIFIC HEAT CAPACITY

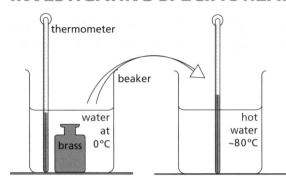

In the example shown, some of the thermal energy stored in the water transfers to thermal energy stored in the brass block. This happens until the temperatures of the block and the water are the same (equilibrium), so:

thermal energy increase of brass = thermal energy decrease of water

$$mass_{brass} \times specific\ heat\ capacity_{brass} \times temperature\ increase_{brass} = mass_{water} \times specific\ heat\ capacity_{water} \times temperature\ decrease_{water}$$

Accurate and safe measurements of the water temperature and time need to be recorded. The specific heat capacity of the brass can then be calculated from the measurements. We know the specific heat capacity of water. The thermal energy wasted to the surroundings is assumed to be zero. If the wasted energy is not zero, the value calculated will not be valid.

Physics practicals

INVESTIGATING IV CHARACTERISTICS

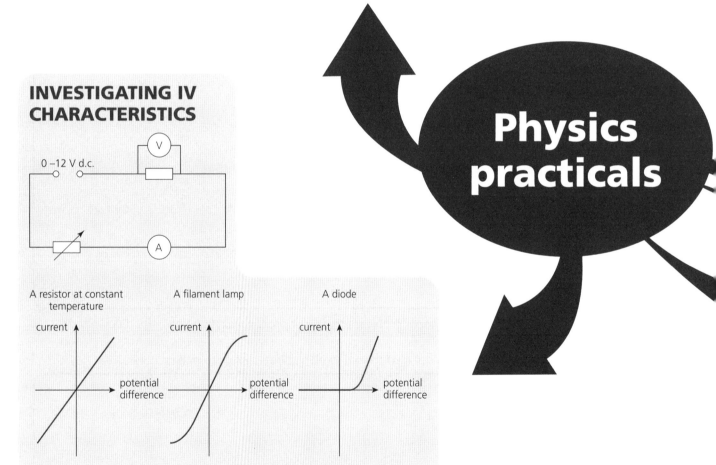

The three variables are: the component being tested; the potential difference, V (that you change); and the current, I (that you measure). The diagram shows the circuit needed to make the measurements. The potential difference is changed in steps over a range using the variable resistor. Reversing the connectors on the power supply gives negative values. Then use the average values to plot the IV characteristics. A graph of I (y-axis) against V (x-axis) can then be plotted for each component. The IV graphs, called characteristics, should look like the ones shown in the diagrams. Uncertainties can be reduced by switching off the power between readings. Repeat the readings and find the mean (average). Then use the mean values to plot the IV characteristics.

USING CIRCUIT DIAGRAMS

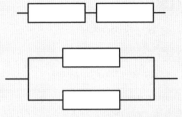

A circuit is set up to measure the potential difference and current for different lengths of a metal wire. The equation:

resistance = potential difference / current

is used to calculate the resistance at each length, and a graph of resistance (*y*-axis) against length (*x*-axis) is plotted. The relationship between length and resistance is determined from the graph. The wire is replaced by two identical resistors in series, and then in parallel. Potential difference and current readings are then taken, and the overall resistance of each combination calculated. These are then compared with the individual resistance of the resistors.

INVESTIGATING DENSITY

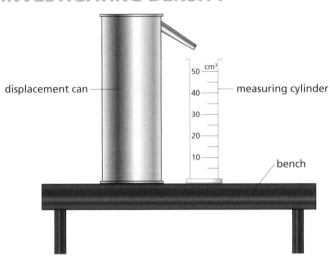

To measure density you need to know the mass and volume of an object. Objects that have a regular shape can be measured to find their volume. Irregularly shaped objects are difficult to measure, so you can find their volume using a displacement can. The volume of the regular object is calculated using:

volume = length × width × height

The measurements can then be used to calculate the density of the objects using:

density = mass / volume

INVESTIGATING ACCELERATION

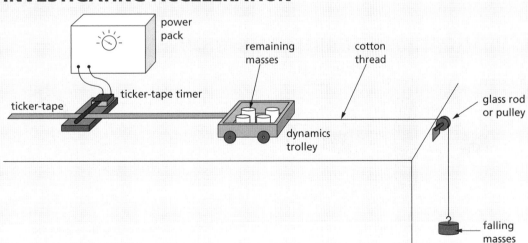

A ticker-tape timer can be used to measure the acceleration of a dynamics trolley. The trolley is pulled by different falling masses that each produce a different pulling force. This investigates the link between force and acceleration. Masses are moved from the trolley to the falling mass stack. This varies the pulling force but keeps the overall mass the same. Adding more masses to the trolley but keeping the pulling force constant, investigates the link between mass and acceleration. The results of the investigation can be compared with Newton's Second Law.

INVESTIGATING THE BEHAVIOUR OF SPRINGS

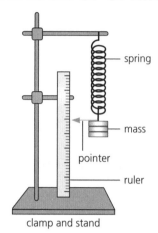

clamp and stand

The independent variable is the force on the spring (produced by changing the mass). The dependent variable is the **extension** of the spring. The main control variable is keeping the spring the same. It is important to measure the extension of the spring NOT the overall length of the spring. Too much force should not be put on the spring, otherwise it will permanently stretch and produce an **uncertainty** in the measurements. A force–extension graph should then be plotted, and the pattern noted.

INVESTIGATING WAVES IN A RIPPLE TANK AND ON A STRING

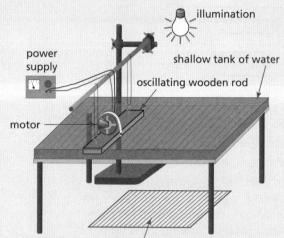

wave patterns on a viewing screen or table

For the ripple tank, a strobe can 'freeze' the pattern of the waves seen on the viewing screen. The frequency of the waves can be read off the strobe and the wavelength is measured accurately with a ruler. The speed of the waves can be calculated using:

wave speed = frequency × wavelength

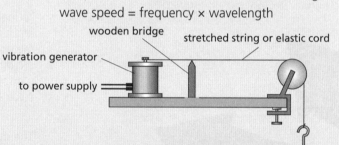

The strobe can also 'freeze' the motion of the stationary waves produced on the vibrating string. The frequency of the waves is measured from the vibration generator power source. The wavelength is measured directly using a ruler.

INVESTIGATING INFRARED RADIATION

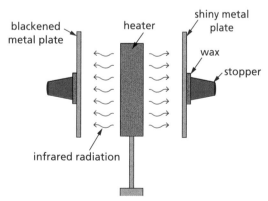

The absorption of infrared radiation by different surfaces is compared using the apparatus shown. The time taken for each stopper to fall off the plates is recorded. The best absorber heats up fastest and the wax melts quickest. The controlled variables are the distance from the heater to the plates and the plate area.

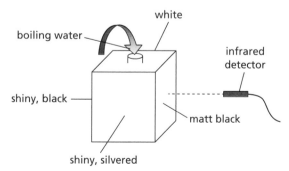

An infrared detector measures the intensity of infrared radiation emitted from each side of the Leslie cube. The experiment must be performed quickly to ensure the water is the same temperature for each reading. The distance between each face and the detector must also be controlled.

PHYSICS SKILLS

Measuring forces – forces are normally measured using a newtonmeter, which is also known as a spring-balance. A force extends a spring, and the position of a pointer attached to the spring is read off a scale to determine the force. Different force ranges can be measured by using newtonmeters with springs with different stiffnesses. Newtonmeters can have a hook on the end to measure tension (pull) forces, or they can have a flat pad to measure compressions (pushes).

Measuring speed – speed requires the measurement of a distance (using a ruler or a tape measure) and the measurement of a time (using a stopwatch). The speed is calculated using the equation:

$$\text{speed} = \frac{\text{distance}}{\text{time}}$$

Instantaneous speed is the speed of an object at any given point during its motion; mean speed is the mean speed over the course of the motion. A 100 m sprinter can run at a maximum instantaneous speed of about 12 m/s, but their mean speed over the whole 100 m is about 9 m/s.

Physics practicals

Calculating from graphs – there are several quantities that can be measured from a graph. The gradient (or slope) of a straight line can be determined by measuring the change in y-axis values of the line and the change in x-axis values of the line. The gradient is determined by:

$$\text{gradient} = \frac{\text{change in } y\text{-axis values}}{\text{change in } x\text{-axis values}}$$

In the graph shown

$$\text{gradient} = \frac{AC}{BC}$$

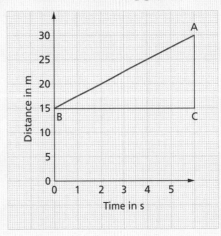

Values can also be read from lines of best fit. It is always best to use a ruler to draw horizontal and vertical lines from the line of best fit. You can then read off the values from the scales on the axes.

Hazards in physics – some of the required practicals in physics involve hazards. The specific heat capacity experiment involves using hot water, which requires the use of safety goggles. Hot objects should be allowed to cool before handling. During the electrical experiments, if mains power supplies are being used, these should be kept away from water taps and sinks. The density experiment has no obvious risks. The acceleration and the spring experiments both involve the use of weights, which could fall onto feet. Feet should be kept out of the way of potential falls. The ripple tank experiment also potentially involves using a mains power supply and water. These should be kept far apart. The infrared experiments involve a mains infrared heater and a Leslie cube, both of which can get very hot. These should be allowed to cool before handling.

Using equations to explain observations – in several of the required practicals, you are asked to investigate the relationship between a dependent and an independent variable that are linked through an equation. Examples of this are:

potential difference = current × resistance

resultant force = mass × acceleration

force applied to a spring = spring constant × extension

wave speed = frequency × wavelength

Each of these can produce a graph with a straight line. If the data values collected form a straight line on the graph, then the equation has been proved.

Session 30 Solving the question

Design an investigation that shows the effect of increasing the mass of an object hung from a spring.

SOLVE

Now you have gathered the relevant information it is really important that you think carefully about how you need to use and present that information. Use the prompts below to help you.

This question asks you to **design** an investigation. This means you need to set out what needs to be done. You should set out how changing the mass hung from a spring affects the extension of the spring.

What equipment would you use for this procedure? You can show this using a diagram. List the range of masses you would use. How will you measure the extension of the spring?

List the steps of the practical procedure.

What are the hazards? What precautions would you take to avoid harm?

What would you expect to happen?

Why would you expect this?

There are additional practice questions with the writing frames for 'Draw', 'Estimate', 'Justify' and 'Sketch' at the end of the book.

Additional writing frames and questions

Now you have completed the sessions, you will have answered questions that use the main command words that can be included in the exams. The following questions will help you practice using each of those command words and revisit the topics from the sessions. You can work on the questions in any order. Your memory will be strengthened most if you mix up the command words and topics that you work on. Each command word has a writing frame. Use the writing frames to help you when you're practising answering exam-style questions. After each writing frame is a list of questions that you can practise with in the writing frame.

Additional writing frames

Balance

Questions that use the word **balance** as a command word need you to balance a chemical equation. This means you need to change the amount of some substances so there is an equal number of each type of atom on both sides of the arrow. Chemical equations should have the state symbols of the substances written next to the substance: (s), (l), (g) and (aq) – always check that you have added these.

Think about what you already know about the reaction. Scribble what you know, particularly any clues about what is reacting and what is being produced. Use chemical formulae where you can.

Now think about the information you need to balance this specific equation. Some of what you noted down before may not be relevant. You only need to think about the atoms in the reaction. The names of substances are not needed.

Now look at the atoms on both sides of the arrow. Count how many there are of each type.

If there are more on one side of the arrow than the other, you will need to increase the amount. You do this by putting a number in *front* of the atoms. If the atom is combined with others in a compound, you will need to increase the amount of the whole thing. This means that balancing one atom unbalances another.

Now check again for any unbalancing caused by the numbers you have added and increase the amount where needed.

Do a final check to make sure:

– you have the same number of atoms on both sides of the arrow

– that the number is in front of the chemical formula

– that the number is big and is written above the line.

You can use the writing frame with any exam-style questions you want to practise that use the same command word. Below are some questions you can try to get you started.

Session 14

Copper reacts with oxygen to make copper oxide.

a) Write the word equation for the reaction.   **[2 marks]**

b) The unbalanced symbol equation for the reaction is:

$$\boxed{}Cu\boxed{} + O_2\boxed{} \rightarrow \boxed{}CuO\boxed{}$$

   i) Add state symbols to the equation.   **[3 marks]**

   ii) Balance the equation.   **[2 marks]**

Session 20

Sodium carbonate reacts with hydrochloric acid to form sodium chloride, water and carbon dioxide.

a) Write the word equation for the reaction.   **[2 marks]**

b) The unbalanced formula equation for the reaction is given below.
Balance the formula equation.   **[2 marks]**

$$Na_2CO_3(s) + \boxed{}HCl(aq) \rightarrow \boxed{}NaCl(aq) + H_2O(l) + CO_2(g)$$

Calculate

Questions that use the word **calculate** need you to use the numbers in the question to work out a value. This often means you need to recall, rearrange or use an equation, and always means you need to use the right units.

Think about what you already know about how the numbers in the question are related. Scribble what you know, particularly any equations and units you can remember.

Now think about the information you need for this specific calculation. Some of what you noted down before may not be relevant. You only need to think about one equation in each question so any other equations are not needed. Check to see if you need to rearrange the equation.

You now need to write down the equation you are going to use. Make sure you have made a note of the units for each part of the equation.

Then put the numbers from the question into the equation.

Now do the calculation on your calculator. Be very careful as it is very easy to put in the wrong numbers. If in doubt, work it out in sections and write down the answer to each section. Always check your calculation on the calculator twice.

Do a final check to make sure your answer is sensible. Look again at the question and ask yourself if the number you have calculated is around what you would expect. It is very easy to put in an extra 0 on the calculator. You might not be able to guess the exact number, but you should spot if you are a long way out.

Finally, check to see if you need to add the units to your answer.

You can use the writing frame with any exam-style questions you want to practise that use the same command word. Below are some questions you can try to get you started.

Session 1

A light microscope can be used to view plant cells such as onion cells, and animal cells such as cheek cells.

a) A student viewed a slide of cheek cells using a light microscope and made the drawing shown below.

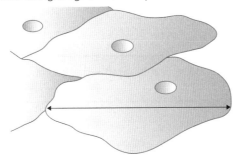

    i) Label three features of the cheek cell shown in the diagram. **[3 marks]**

    ii) In real life the cell shown measures 40 µm across.

        Calculate the magnification of the cell in the image. **[3 marks]**

        Measure the width of the cell in millimetres (mm):

        Convert your measurement to micrometres (µm):

        Calculate the magnification.

b) Describe how to make and view an onion tissue specimen on a light microscope. Include a list of equipment and refer to any safety concerns in your answer. **[6 marks]**

Session 14

Copper is a valuable metal. Copper is found as copper carbonate in some rocks. The copper carbonate reacts with carbon to make copper and carbon dioxide.

a) State why the reaction of copper carbonate with carbon is a chemical change. **[1 mark]**

b) 247 kg of copper carbonate reacts with 12 kg of carbon to form 127 kg of copper.

    i) Write the word equation for this reaction. **[2 marks]**

    ii) Calculate the mass of carbon dioxide that would be produced in this reaction. **[2 marks]**

Session 24

The graphs show the distance travelled by a skateboarder over three skate runs in a skate park.

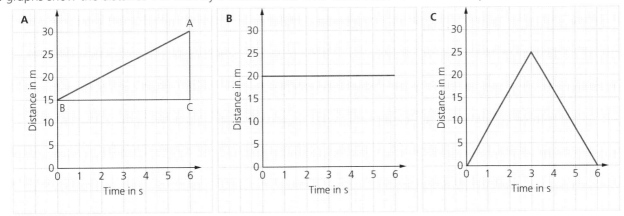

a) Describe the skate runs shown in each graph. **[3 marks]**

b) Calculate the speed of the skateboarder:

    i) in graph A, between 0 and 6 seconds **[1 mark]**

    ii) in graph C, between 0 and 3 seconds. **[1 mark]**

c) Identify when the skateboarder was moving fastest forwards, away from an observer at the origin (0,0). **[1 mark]**

d) Identify when the skateboarder was going backwards. **[1 mark]**

Compare

This question asks you to **compare**. This means that you need to write about similarities and differences between two things. What are you comparing?

Decide which topic the question is asking about. Think about what you already know about this topic. Scribble what you know, particularly key words.

Now think about the information you need for this specific question. Not everything you noted down before may be relevant.

How are the two things similar?

How are they different?

Now write down the whole answer: write out what is similar and what is different. Make sure you write about both things equally. Do not just write about one, or much more about one than the other. Finally, write a conclusion based on your comparisons.

You can use the writing frame with any exam-style questions you want to practise that use the same command word. Below are some questions you can try to get you started.

Session 2

When you exercise your muscle cells use more oxygen than when you are not exercising.

a) Name the structure in the muscle cell that uses oxygen in respiration. **[1 mark]**

b) State the word equation for respiration. **[1 mark]**

c) Compare the diffusion of oxygen into muscle cells when you are exercising with when you are not exercising. **[4 marks]**

Session 4

The diagram shows the human heart.

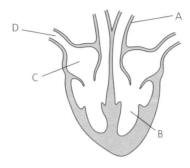

a) Identify labels A to D. **[4 marks]**

b) Compare the two sides of the heart, stating their similarities and differences. **[4 marks]**

Sessions 5 and 6

The internal environment of the body needs to be carefully controlled to prevent harm and disease.

Compare how the reflex response to touching something hot is different from the hormonal control of blood sugar. **[6 marks]**

Session 10

Farmers have used selective breeding for centuries to change the characteristics of organisms used for food. In modern times genetic engineering does the same thing.

Compare selective breeding and genetic engineering in farming. **[4 marks]**

Session 26

Compare the production of electricity by a nuclear power station to the production of electricity by a wind turbine. **[3 marks]**

Session 29

The diagram shows a helium atom.

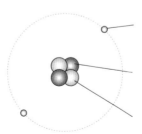

a) Label the three subatomic particles. **[2 marks]**

b) Compare the relative masses and the charge of each part of the atom you have labelled. **[2 marks]**

c) Explain how an alpha particle is different from a helium atom. **[2 marks]**

Define

Questions with the command word **define** need you to write down exactly what something means. Work out which word or phrase you need to define.

Scribble down your thoughts and use the knowledge organiser to fill in any gaps in your knowledge.

Write down any scientific words you could include in your answer.

Now write out your final definition of the word or phrase. Try to use scientific words and write in full sentences.

You can use the writing frame with any exam-style questions you want to practise that use the same command word. Below are some questions you can try to get you started.

Session 6

The menstrual cycle results in the release of an egg. If the egg is fertilised by a sperm cell then a pregnancy will start. A pregnancy can be prevented in two ways:

- hormonal methods
- non-hormonal methods.

The table below lists some common hormonal and non-hormonal methods of contraception.

Hormonal methods	Non-hormonal methods
Daily oral contraceptives – contain hormones that inhibit the production of FSH so that no eggs mature.	Condoms – prevent the sperm reaching an egg.
Skin patches or implants – slowly release the hormone progesterone which inhibits the maturation of eggs for several months or years.	Spermicidal creams or gel – kill or disable sperm.
Intrauterine device (coil) – slowly releases the hormone progesterone which inhibits the maturation of eggs for several months or years.	Sterilisation surgery – the fallopian tubes in females are blocked, preventing the release of eggs, or the tubes from the testes are blocked in males, preventing the release of sperm.

Define the terms:

a) hormonal method of contraception [1 mark]

b) non-hormonal method of contraception. [1 mark]

Session 7

Malaria is spread by mosquitos which act as a vector. A vector is an organism that transfers a pathogen from one organism to another. The mosquito picks up the pathogen from one person when it bites them. The mosquito then passes the pathogen to the next person it bites. Mosquitos breed in water, preferring to lay their eggs in stagnant water.

a) Define the term 'pathogen'. [1 mark]

b) Frank lives in a village next to some stagnant water ponds. He suggests draining the ponds as a way of controlling the spread of malaria. Justify Frank's suggestion as a way to reduce the number of malaria infections in his village. [2 marks]

c) Describe **one** more way that the spread of malaria can be controlled. [1 mark]

Session 14

During boiling, liquid water molecules undergo a physical change into water vapour molecules. During electrolysis, water molecules undergo a chemical change into oxygen and hydrogen gas. Define the terms:

a) physical change [1 mark]

b) chemical change. [1 mark]

Session 27

The resistance of components in an electrical circuit is an important property that affects the currents and potential differences within the circuit.

a) Define resistance. [1 mark]

b) The table shows three ways to change the resistance of electrical components. Tick the correct box to identify the correct change on the resistance of the component. [3 marks]

	increases the resistance	decreases the resistance
Increasing the length of a metal wire …		
Increasing the temperature of a thermistor …		
Increasing the light intensity on a light dependent resistor (LDR) …		

Session 29

The diagram shows the nucleus of two different atoms of hydrogen. The two types of hydrogen are isotopes.

hydrogen-1 nucleus
(1 proton)

hydrogen-2 (deuterium) nucleus
(1 proton plus 1 neutron)

Use information in the diagram to define the term 'isotope'. [2 marks]

Additional writing frames

Describe

Questions using the command word **describe** need you to remember some facts or a process and write it down clearly. You may also be asked to describe what is shown in a graph or diagram. Describe questions ask you to write about what is happening, whereas explain questions ask you to say why something happens.

Look at the information in the question to find clues about the topic, procedure or graph you need to write about.

Think about what you already know about the topic or what is described by the graph in the question. Look for any key words and think about what they mean. Scribble what you know, particularly any definitions or processes.

Now think about the information you need for this specific question. Some of what you have written may not be helpful. Remember that you just need to stick to the facts to say what happens, not why it is happening.

Next, put your thoughts in a sensible order. If you are describing a graph, write some key data points from the graph here. Jot down some describing words that might be useful. Is the graph showing an increase or decrease in values? What key words could you use to describe the process you are writing about?

Now write your final answer. Make sure you include any values from the graph. Use scientific words in your answer.

Do a final check to make sure you have fully answered the question. Is everything the question asked about covered in your answer?

You can use the writing frame with any exam-style questions you want to practise that use the same command word. Below are some questions you can try to get you started.

Session 2

Humans can get very ill if they drink a large amount of water in a short time.

a) Describe how the water travels through the cells of the gut to the blood. **[2 marks]**

b) Explain why blood cells become damaged when lots of water enters the bloodstream in a short time. **[3 marks]**

Session 3

The function of the lungs is to absorb oxygen from the air we breathe in. Describe how the lungs are adapted to absorb oxygen by diffusion. **[4 marks]**

Session 8

The growth of the human population is changing the planet.

a) Describe how burning rainforests to make space for crop production affects biodiversity. **[3 marks]**

b) Describe how burning fossil fuels for transport causes climate change. **[2 marks]**

Session 13

All the elements in Group 1 of the periodic table react with water in a similar way.

a) The table below shows some properties of Group 1 elements. Identify which properties are correct (✓) and which are incorrect (✗).

Property of Group 1 element	Correct (✓)	Incorrect (✗)
All are metals.		
All are shiny when cut with a knife.		
All react with water producing carbon dioxide gas.		
All react with water producing an alkaline solution.		

[4 marks]

b) i) Describe how the radius of the atoms of the elements of Group 1 changes as you go down the group. **[1 mark]**

ii) Describe how the reactivity of the atoms of the elements of Group 1 changes as you go down the group. **[1 mark]**

iii) Explain why the reactivity of Group 1 metals changes as you go down the group. Make sure you write about electron structure and the effect of the changing radius of the atoms. **[3 marks]**

Session 26

A fridge uses electrical energy to cool food.

a) Describe the energy transfers involved in:

i) the fridge using electrical energy to turn the light on when the fridge door is opened **[2 marks]**

ii) the fridge cooling the food. **[2 marks]**

b) The fridge contains ice.

i) Describe how the store of kinetic energy of the particles in the ice is different from the store of kinetic energy in the particles of liquid water. **[1 mark]**

ii) Compare the movement of the particles in the ice and the liquid water. **[2 marks]**

Design

This question asks you to **design**. This means that you need to set out how a practical procedure will be done. Decide which practical method the question is asking about. Think about what you already know about this method. Scribble what you know, particularly the names of apparatus or anything you can measure.

Now think about the information you need for this specific question. Not everything you noted down before may be relevant.

Now you have gathered the relevant information think carefully about how you are going to organise it. Use the prompts below to help you.

Decide the equipment you will need and write a list or draw a diagram.

What will you change in the investigation? This is the independent variable.

How will you keep everything else the same? These are the control variables. List them here.

Describe how you will keep the control variables constant.

What will you measure? This is the dependent variable.

How will you measure it?

How many times will you measure it? Think about how often you need to take a measurement.

Are there any risks? How will you manage the risk?

Now write down the whole method: you can use bullet points and labelled diagrams. Make sure you write about the whole method, from setting it up to how you will record the results.

You can use the writing frame with any exam-style questions you want to practise that use the same command word. Below are some questions you can try to get you started.

Session 11
Design an investigation into the effect of caffeine on reaction time. **[6 marks]**

Session 22
Design an investigation that will allow you to find out which of three pens has the most different colours in its ink. **[6 marks]**

Additional writing frames

Determine

For **determine** questions you need to use what you are given in the question. Think about what you already know about the topic and what the information in the question means. Look for any key words and think about what they mean. Scribble what you know, including any useful information from the question.

Now think about the information you need for this specific question. Remember you need to use the information given so some of what you noted down before may not be relevant.

Look again at the question and try to find clues to help you answer it. Highlight the information that you need to use to work out the answer. This could be words or values.

Now bring together your notes above to write your final answer. You must show any working that you have used at each stage of your answer. Check to make sure everything asked about is covered in your answer. If you have done a calculation, check it again using your calculator.

You can use the writing frame with any exam-style questions you want to practise that use the same command word. Below are some questions you can try to get you started.

Session 10

Cystic fibrosis is an inherited condition caused by a recessive allele. A mother who has cystic fibrosis and a father who is a carrier decide to have a child.

a) Complete the Punnett square to give the four possible cystic fibrosis genotypes of the offspring.

Use F for the dominant allele and f for the recessive allele. **[2 marks]**

b) Determine the probability of the child being born with cystic fibrosis. **[1 mark]**

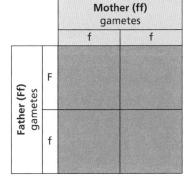

Session 12

The table below shows three key properties of covalent molecules.

Covalent molecule	Molecular mass (g/mol)	Melting point (°C)	Boiling point (°C)
Naphthalene	128	80	218
Water	18	0	100
Hydrogen	2	−259	−252

a) Explain how the molecular mass of a covalent molecule affects its boiling point and melting point. **[2 marks]**

b) At room temperature (25 °C), determine which substance is a

   i) solid **[1 mark]**

   ii) liquid **[1 mark]**

   iii) gas. **[1 mark]**

Session 13

The diagram shows a section of the periodic table.

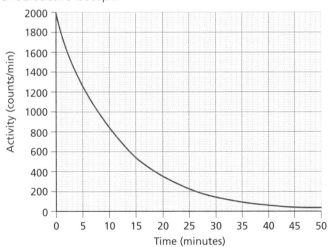

a) Determine the number of:

   i) protons in an atom of argon [1 mark]

   ii) electrons in an atom of sulfur [1 mark]

   iii) electrons found in the outer shell of a fluorine atom [1 mark]

   iv) neutrons in an atom of oxygen. [1 mark]

b) Identify the chemical symbol for tin. [1 mark]

Session 29

The graph shows the decay of a radioactive isotope.

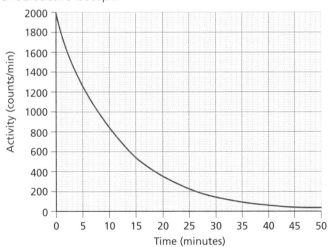

Determine the half-life of the isotope. [2 marks]

Draw

This question needs you to use your science to make or add to a diagram. Look for clues about the topic in the question and think about what you already know about it. Scribble down your thoughts and use the knowledge organiser to fill in any gaps in your knowledge.

Now look at the space or diagram you need to **draw** on. Make sure you know what it is showing and what you need to draw. You should try to fill most of the space you are given for a diagram.

Write some notes here about any labels you need to include. Remember to use scientific words. You could draw a rough sketch here too.

Now complete your drawing. This may be adding to a diagram or drawing one of your own. Use this checklist as a guide:

- Draw in pencil.
- Use a ruler for lines.
- Use clear labels with lines that touch what they describe.
- Do not use any shading (unless asked to in the question).
- Draw only in 2D.

You can use the writing frame with any exam-style questions you want to practise that use the same command word. Below are some questions you can try to get you started.

Session 15

An ionic bond is formed when metals and non-metals react together during a chemical change.

a) Describe how an ionic bond is formed between sodium and chlorine. **[3 marks]**

b) Draw a diagram to show how the sodium atom becomes a sodium ion. **[3 marks]**

Session 15

Hydrogen atoms bond with an oxygen atom to make a water molecule by sharing pairs of electrons.

Draw a dot and cross diagram to show the covalent bonding in a water molecule. **[3 marks]**

Session 22

Draw a diagram to show how you would prepare a solid, dry sample of copper sulfate from a solution of copper sulfate. **[2 marks]**

Session 22

Sodium thiosulfate is a colourless solution. When it reacts with a dilute acid it produces a precipitate that blocks a cross on a piece of paper below.

a) You are investigating the effect of temperature on the rate of the reaction between sodium thiosulfate solution and dilute hydrochloric acid. Draw a labelled diagram of the apparatus you would use. **[3 marks]**

b) Explain why the reaction is faster when the temperature is higher. **[2 marks]**

Session 27

The diagram shows a resistor connected in a series circuit.

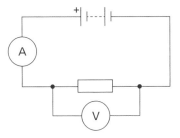

a) Draw a circuit diagram for a battery and three lamps connected in

   i) series **[2 marks]**

   ii) parallel. **[2 marks]**

b) Identify a place on each diagram where a switch would control all three lamps at the same time. Mark these positions with an X. **[2 marks]**

Session 30

Draw a diagram to show how a Leslie cube can be used to investigate how different surfaces emit infrared radiation. **[3 marks]**

Estimate

Estimate questions need you to use what you know about a topic or practical procedure to guess a value. Think about what you already know about the topic or process you would use to estimate a value. Scribble down your ideas.

Now think about the information you need. Some of what you noted down before may not be relevant. You are most interested in any methods you could use to estimate the value.

You now need to write down how you would come up with your estimate. Make sure that you include all your ideas and state your estimate at the end.

If you do need to give a value, make sure you include units. Double-check that the answer is around what you would expect and not a lot bigger or smaller.

You can use the writing frame with any exam-style questions you want to practise that use the same command word. Below are some questions you can try to get you started.

Session 23

The diagram shows an aeroplane during a flight. The plane's engines produce thrust, which pushes the plane through the air.

Estimate the value of air resistance when the value of the thrust from the engine is 500 000 N:

a) if the plane is moving at a steady speed **[1 mark]**

b) if the plane is slowing down **[1 mark]**

c) if the plane is getting faster. **[1 mark]**

Session 25

The diagram shows a wave.

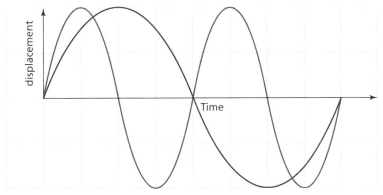

a) The frequency of the red wave is 4 Hz. Estimate the frequency of the blue wave. **[1 mark]**

b) Label the diagram to show:

i) the wavelength of one of the waves **[1 mark]**

ii) the amplitude of one of the waves. **[1 mark]**

Session 30

A toy car accelerates at 2 m/s² when a resultant force of 2 N acts it.

Estimate the acceleration of the toy car when the resultant force acting on the toy car increases to 4 N.

[1 mark]

Session 30

The resistance of a 30 cm length of wire is 5 Ω.

Estimate the resistance of a wire with a length of 120 cm. **[1 mark]**

Evaluate

This question asks you to **evaluate**. This means that you need to consider the positive and negative points about the information in order to decide. You need to use what you know alongside the information given in the questions to make a judgement.

Look at what topic the question is asking about. Think about what you already know about this topic and look for helpful information in the question. Scribble what you know and which parts of the question you can use.

Now think about the information that will help you make your judgement. Not everything you noted down before may be relevant.

Now think about what you need to do with the information you have to answer this question. What are the positive points about the information?

What are the negative points about the information?

What scientific words do you know that you could include in your answer?

Now write down the whole answer:

- Write something about both the positives and negatives. Make sure you write about both equally – do not just write about one, or much more about one than the other.
- Make sure you come to a judgement and use the evidence to state what your opinion is.

You can use the writing frame with any exam-style questions you want to practise that use the same command word. Below are some questions you can try to get you started.

Session 4

Scientists suggest that there is a link between smoking and lung cancer. The graphs show trends in the number of cigarettes smoked and the cases of lung cancer for men and women from 1900 to 1990.

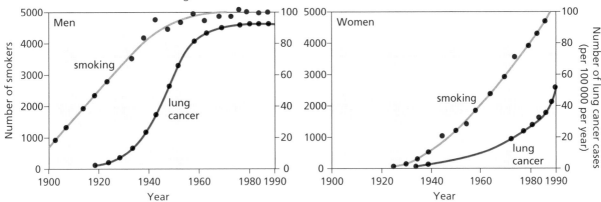

a) Describe how the number of smokers has changed over time for

   i) men   **[1 mark]**

   ii) women.   **[1 mark]**

b) **i)** Evaluate the evidence in the graphs for the argument that smoking causes cancer.   **[3 marks]**

   ii) Suggest why there is a link between smoking and cancer.   **[2 marks]**

Session 12

The diagram represents a model of the particles in a solid, a liquid and a gas.

a) Evaluate the limitations of the model.   **[2 marks]**

b) Compare the structure of a solid to the structure of a gas.   **[2 marks]**

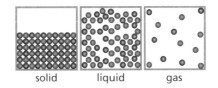

solid   liquid   gas

Session 16

Graphite and copper both have free electrons in their structure.

a) Suggest why both substances conduct electricity.   **[2 marks]**

b) Evaluate the suitability of each element for use in a wire in an electrical circuit.   **[3 marks]**

Session 19

Aluminium is extracted from a molten mixture by electrolysis.

a) Give a reason why aluminium cannot be extracted by heating with carbon, which is cheaper.   **[1 mark]**

b) State why the aluminium mixture has to be a liquid for electrolysis to work.   **[2 marks]**

c) Aluminium can be made by recycling used aluminium.

   Evaluate the production of aluminium by electrolysis and aluminium recycling.   **[4 marks]**

Session 20

Most soils have a pH value between 4 and 10. It is difficult to grow plants in soil that is too acidic. A farmer uses a soil additive made from limestone to neutralise their acidic soil.

a) **i)** Suggest a pH value for the farmer's acidic soil.   **[1 mark]**

   ii) Suggest a pH value for the soil after treatment if it is successfully neutralised.   **[1 mark]**

b) A model of this reaction can be made in the laboratory by adding limestone chips to hydrochloric acid.

   i) Evaluate this reaction as a model for the neutralisation of soil by a limestone additive.   **[1 mark]**

   ii) Describe what you would observe during the reaction.   **[2 marks]**

   iii) The reaction took place in a beaker on an electronic balance.

     Describe what would happen to the reading on the electronic balance during the reaction.   **[1 mark]**

   iv) Explain your answer to **iii)**.   **[1 mark]**

   v) Describe **two** ways by which the rate of reaction could be increased.   **[2 marks]**

Explain

Questions using the command word **explain** need you to use your science to say why something is happening.

Look at the information in the question to find clues about the topic or results of an investigation you need to write about.

Think about what you already know about the topic or investigation that the question is asking about. Look for any key words and think about what they mean. Scribble what you know, particularly any important ideas that are related to this topic.

Now think about the information you need for this specific question. Some of what you have written may not be helpful. Remember that you need to say why something is happening so facts on their own will not be enough.

Next, try to link ideas that answer the question, and write these ideas down in order. Make sure you are using scientific words in the right way.

Do a final check to make sure you have fully answered the question. Is everything the question asked about in your answer?

You can use the writing frame with any exam-style questions you want to practise that use the same command word. Below are some questions you can try to get you started.

Session 1

Pancreas cells make proteins called enzymes that break down food in the gut during digestion.
Pancreas cells have a lot of ribosomes.

a) State why ribosomes cannot be seen under the light microscope. **[1 mark]**

b) Explain how pancreas cells are adapted to their function. **[3 marks]**

Session 3

The cells in the leaf of a plant carry out photosynthesis, which requires carbon dioxide. The diagram shows a cross-section of a leaf.

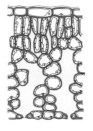

a) Explain how the leaf is adapted for the movement of carbon dioxide by diffusion. **[3 marks]**

b) i) State where in the cell photosynthesis happens. **[1 mark]**

ii) State the word equation for photosynthesis. **[2 marks]**

Session 7

A potential vaccine for HIV contains only a part of the virus.

a) Explain how introducing a part of the virus into the body would protect people from HIV infection. **[4 marks]**

b) State the advantage of only using a part of the virus in the vaccine. **[1 mark]**

Session 20

The diagram shows the reaction between magnesium and a solution of iron(II) sulfate.

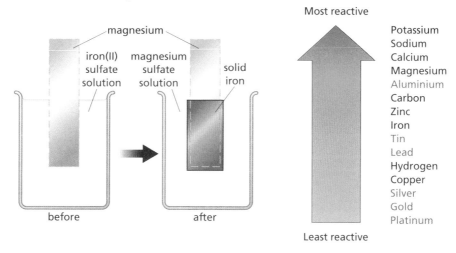

Use the reactivity series to explain why magnesium can displace iron from a solution of iron(II) sulfate. **[2 marks]**

Session 24

The stopping distance of a vehicle depends on the driver, the conditions (e.g. weather, road surface) and the car.

a) Complete the table using (✓) to identify the factors that affect thinking distance, braking distance or both. **[5 marks]**

Factor	Thinking distance	Braking distance	Both
Speed of car			
Water on road			
Driver's tiredness			
Driver's alcohol consumption			
Condition of car's brakes			

b) When travelling at the same speed, a fully loaded van takes longer to stop than the same van when it is empty. Use what you know about energy and forces to explain why. **[3 marks]**

Additional writing frames

Justify

To answer a **justify** question you need to use only what you are given in the question, so look carefully at the information in the question.

Think about what you already know about the topic and what the question is asking you. Make a note of the main idea and any key words that you know and are in the question.

Now think about the information you need for this specific question. Remember you need to use only the information given so some of what you noted down before may not be relevant.

Highlight the key bits of information in the question.

Next, write your full answer using what is in the question. Do a final check to make sure everything asked about is covered in your answer.

You can use the writing frame with any exam-style questions you want to practise that use the same command word. Below are some questions you can try to get you started.

Session 14

Calcium carbonate is a white solid. Some students use a pestle and mortar to break up a large piece of calcium carbonate. They then heat a small piece in the flame of a Bunsen burner. They weigh the piece of calcium carbonate before and after heating, and find that it had less mass after heating.

a) **i)** Identify which part of the process was a physical change and which was a chemical change. [1 mark]

  ii) Justify your answer to **a) i)**. [1 mark]

b) When calcium carbonate is heated it changes into calcium oxide and carbon dioxide.

  i) Write a word equation for this reaction. [2 marks]

  ii) The formula for calcium carbonate is $CaCO_3$. Determine how many of each atom there are in the molecule. [3 marks]

Atom	Number in $CaCO_3$
Ca	
C	
O	

Session 16

A new type of ionic liquid is being developed for use in batteries. The substance melts just below room temperature.

a) Determine what state the substance would be at 2 °C. **[1 mark]**

b) Justify why the substance needs to be in the liquid state for it to conduct electricity. **[3 marks]**

Session 17

The diagram shows how the atoms rearrange when magnesium reacts with oxygen.

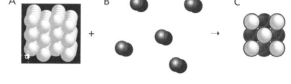

a) Determine which diagram(s):

  i) show an element

  ii) show a compound.

b) Justify your reasons for your answers to **a) i)** and **ii)**.

Session 21

Copper sulfate and water can react reversibly. The equation for this reaction is:

$$\text{hydrated copper sulfate (blue)} \underset{\text{exothermic}}{\overset{\text{endothermic}}{\rightleftharpoons}} \text{anhydrous copper sulfate (white)} + \text{water}$$

a) When blue hydrated copper sulfate is heated it becomes white anhydrous copper sulfate and water is given off. State how you know that this is a chemical reaction. **[1 mark]**

b) Priya added a spatula of anhydrous copper sulfate to a test tube half full of water. She first measured the temperature of the water to be 16 °C. One minute after adding the anhydrous copper sulfate, she measured the temperature of the solution and found it to be 25 °C. Priya concluded that the reaction was an example of an exothermic reaction. Justify Priya's answer. **[1 mark]**

Session 29

The diagram shows the penetrating power of three different forms of nuclear radiation.

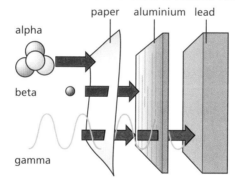

Use information in the diagram to justify why sources of gamma radiation are kept in lead-lined containers. **[2 marks]**

Session 30

The diagram shows the set-up of an experiment to measure the speed of waves. A student uses a ruler and a stopwatch to take measurements of the wave patterns observed on the viewing screen.

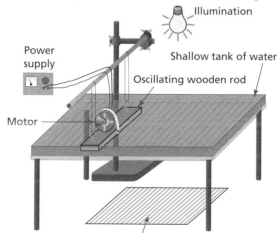

Justify the use of the lamp in the set-up of the ripple tank in the diagram. **[2 marks]**

Plan

This question needs you to write a method for an investigation. Look at what the question is trying to find out and think about when you have done or seen the practical. Scribble what you know, particularly the names of apparatus, any ways of making changes or anything you can measure.

Now think about the information you need for this specific question. Not everything you noted down before may be relevant.

Next, think about how you will explain how to do the method.

Decide the equipment you will need and write a list or draw a labelled diagram.

What will you change in the investigation? This is the independent variable.

How will you keep everything else the same? These are the control variables.

What will you measure? This is the dependent variable.

How will you measure it?

How many times will you repeat the measurements?

Are there any hazards?

How will you manage risk?

How will you analyse the results of your investigation?

Now write down the whole method: you can use bullet points and labelled diagrams. Make sure you write about the whole method, from setting it up to what the results mean.

You can use the writing frame with any exam-style questions you want to practise that use the same command word. Below are some questions you can try to get you started.

Session 11

The enzyme amylase breaks down starch in digestion.

Plan an investigation that shows what happens to the rate of this reaction when the pH is changed. **[6 marks]**

Session 22

Specific heat capacity is the energy needed to raise the temperature of 1 kg of a substance by 1 °C.

Plan an investigation you could use to find out about the different specific heat capacities of four different metal blocks. Each metal block has two holes in – one for a thermometer and one for a heater, as shown in the diagram. **[6 marks]**

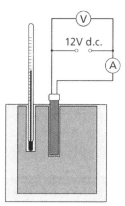

Plot

This question will give you some data that you need to put onto a graph. First find the data you will **plot**. Then look at the graph. You will often have axes drawn for you, but it is a good skill to be able to do this yourself.

Draw the axes if this has not been done for you. The horizontal axis is called the x-axis and runs along the bottom of the graph paper. The vertical axis is called the y-axis and runs up the side of the graph paper. Your axes must be drawn with a ruler and a pencil. The scale on the axis needs to be big enough to fit on all of the data. The scale should include the smallest number to the biggest, with each interval being the same size. Each axis needs a label with units.

The independent variable, or the thing that changed in the investigation goes on the x-axis. The dependent variable, or the thing you measured goes on the y-axis.

Next, you need to decide what type of graph to use. Data where one of the variables is non-numerical does not have a link between each point and should be drawn as a bar chart. An example of non-numerical data is colour. Data where both variables are numerical should be drawn as a line graph. An example is the number of leaves of different lengths. Line graphs are helpful to predict measurements that you did not take in the investigation.

Finally, you need to plot your data on the axes. Use a sharp pencil and a ruler to draw any lines. Make sure you draw a line of best fit if the question asks for it. The line of best-fit:

- shows the trend
- could be a straight line or a curve
- should have an equal number of points on either side of it
- does not include anomalies – odd numbers in the data that do not follow the trend. Anomalies should be circled if they are plotted on a graph, and ignored.

You can use the writing frame with any exam-style questions you want to practise that use the same command word. Below are some questions you can try to get you started.

Session 24

The table shows the velocity of a racing car during a race.

Time (s)	0	1	2	3	4	6	8	10	12
Velocity (m/s)	0	10	20	29	36	50	59	64	64

Plot a velocity–time graph for the data shown in the table. Remember to include labels and a suitable scale. **[6 marks]**

Session 24

The car shown below travels on a journey. During the journey it changes velocity several times.

a) Calculate the acceleration of the car during the following sections of the journey:
   i) stationary at the start of the journey to a velocity of 13 m/s 5 seconds into the journey **[2 marks]**
   ii) and then joining the motorway travelling at a velocity of 13 m/s and speeding up to 30 m/s in 2 seconds. **[2 marks]**

b) The car then travels at 30 m/s for 10 seconds, before braking to a stop in 2 seconds when the car in front skidded.

Plot a graph to show the velocity of the car over the whole journey. **[6 marks]**

Predict

To answer a **predict** question you will need to write what is likely to happen. Look for clues about the topic in the question and think about what you already know about it. Scribble down your thoughts and use the knowledge organiser to fill in any gaps in your knowledge.

Now think about the information you need for this specific question. Not everything you noted down before may be relevant.

What information are you given in the question?

Now use what you know to say what is likely to happen in the situation in the question. Check to see if the question asks you to explain your prediction. If you need to explain, use your science to say why you think your prediction will happen.

You can use the writing frame with any exam-style questions you want to practise that use the same command word. Below are some questions you can try to get you started.

Session 11

a) Predict what will happen to the mass of a sample of beetroot cells when they are placed in:
   i) a concentrated sugar solution [1 mark]
   ii) pure water [1 mark]
   iii) a sugar solution that is the same concentration as the beetroot cells. [1 mark]
b) Give a reason for each of your predictions. [3 marks]

Session 11

The table of nutritional information on the side of a box of eggs gives the following information:

Typical values	100 g contains	Per average egg	% based on Reference Intake for Average Adult
Energy	598 kJ	280 kJ	3
Fat	9.6 g	4.5 g	6
Starch	zero	zero	zero
Sugar	zero	zero	zero
Protein	14.1 g	6.6 g	13
Salt	0.38 g	0.18 g	3

A student performs a series of food tests on a sample of one of the eggs. A table of their results is shown below. Predict the results of each test. Put a tick (✔) in the outcome box if there is a positive result and put a cross (✗) if there is a negative result. [4 marks]

Food test	Outcome of the test (positive = ✓ / negative = ✗)
Benedict's test for sugars	
Iodine test for starch	
Biuret test for protein	
Ethanol test for lipids	

Session 12

Water is an unusual substance because solid water is less dense than liquid water. This means that ice floats on top of liquid water.

a) Explain why water is unusual compared to other substances. Write about the density of solids compared to a liquid of the same substance in your answer. [2 marks]
b) The temperature of the planet is increasing. Ice that usually floats on top of the sea is melting faster than expected.
   i) Describe the effect of rising temperatures on sea levels. [2 marks]
   ii) Predict two effects of rising sea levels on habitats. [2 marks]

Session 28

A student performs an experiment by making an electromagnet from a coil of wire connected to a switch and a power supply. The switch is turned on and metal paper clips are hung from the coil. The student continues to add paper clips until no more will attach to the electromagnet. He records that 12 paper clips can be supported by the electromagnet.

Predict what will happen to the number of paper clips that can be supported when he makes the following changes to the experiment:

a) He doubles the current flowing through the coil. [1 mark]
b) He slides an iron nail through the middle of the length of the coil. [1 mark]

Additional writing frames

Show

Questions with **show** as the command word need you to give clear evidence for your conclusion.

Look at the information in the question to find clues about the topic. This may include results of an investigation you need to write about.

Think about what you already know about the topic or investigation that the question is asking about. Look for any key words and think about what they mean. Scribble what you know, particularly any powerful ideas that are related to this topic.

Now think about the information you need for this specific question. Some of what you have written will not be helpful – remember that you need to have clearly structured evidence in your answer, and it is easy to get off-track.

Next, try to link ideas that answer the question, and write these ideas down in order, or draw a diagram of your ideas. Make sure you use scientific vocabulary in the right way and that you include a final conclusion based on what you have written.

Do a final check to make sure you have fully answered the question. Is everything the question asked about in your answer and is there a clear conclusion, backed up by evidence?

You can use the writing frame with any exam-style questions you want to practise that use the same command word. Below are some questions you can try to get you started.

Session 5

Show how reflex actions protect humans from harm. **[6 marks]**

Session 9

Sex cells are produced during the process of meiosis.

a) State where meiosis happens in human
  i) males **[1 mark]**
  ii) females. **[1 mark]**
b) Describe how meiosis results in egg cells and sperm cells that are genetically different from each other. **[1 mark]**
c) All female egg cells contain the XX chromosome combination. All sperm cells contain the XY chromosome combination. The chromosome combination of two parents is shown in the diagram below.

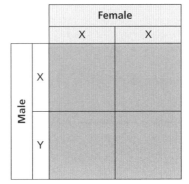

  i) Complete the diagram. **[2 marks]**
  ii) Use the information in the diagram to show how parents have a 50% chance of a female or male embryo being formed. **[2 marks]**

Session 22

Show how you would find the density of a cube of copper. **[4 marks]**

Session 22

Show how you would find the density of an irregular-shaped piece of coal. **[6 marks]**

Session 25

The diagram shows the electromagnetic spectrum with some parts of the spectrum missing.

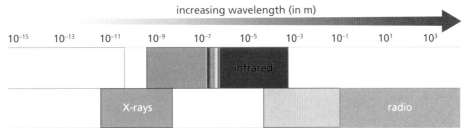

a) Show which waves are missing by writing them in the blank coloured boxes on the diagram. **[3 marks]**
b) Give a use for:
  i) X-rays **[1 mark]**
  ii) visible light **[1 mark]**
  iii) infrared **[1 mark]**
  iv) radio waves. **[1 mark]**

Sketch

This question needs you to use your science to draw approximately – to **sketch**. Look for clues about the topic in the question and think about what you already know about the idea. Scribble down your thoughts and use the knowledge organiser to fill in any gaps in your knowledge.

Now look at the space you need to use and draw an outline. Make sure it is big enough to read, but does not go over the lines at the edge of the paper. Now fill in the details and add labels. Draw in pencil and use a ruler for lines. Make sure you use clear labels with lines that touch what they describe. Do not use any shading unless specified in the question, and only draw in 2D.

I have:

- [ ] drawn in pencil
- [ ] labelled key features with lines from the label that touch what they describe
- [ ] used a ruler for straight lines
- [ ] drawn a diagram that fills the available space
- [ ] not used shading or colouring in
- [ ] drawn only in 2D.

You can use the writing frame with any exam-style questions you want to practise that use the same command word. Below are some questions you can try to get you started.

Session 11

Sketch diagrams to show the method you would use to test a food for:

a) starch [1 mark]

b) sugar [1 mark]

c) protein [1 mark]

d) lipids. [1 mark]

Session 22

a) Sketch a diagram to show how you would set up the electrolysis of sulfuric acid to produce hydrogen gas, chlorine gas and sodium hydroxide solution. [4 marks]

b) Describe the test you would use to show that hydrogen has been produced successfully. [2 marks]

Session 22

Drinking water is a precious resource in many areas of the world. In some parts of the world there is not enough fresh water, so salt is removed from sea water to make it suitable for drinking. This is an expensive process.

a) Evaluate the use of water purification using salt water in less industrialised countries. [2 marks]

b) Sketch a diagram to show you could remove the sand from a beaker of sandy water. [2 marks]

Session 30

On the axes below, sketch a graph to show how the current through a lamp changes when the potential difference across a lamp is changed.

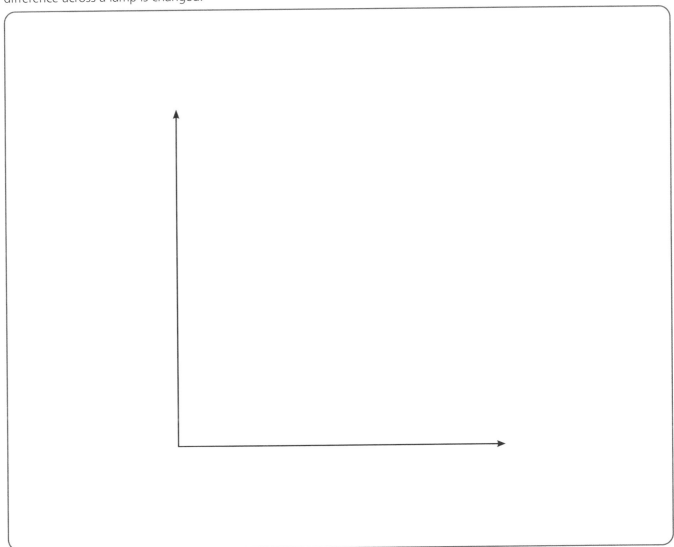

Suggest

Questions using the command word **suggest** mean you need to apply what you know to a new situation.

Look at the information in the question to find clues about the topic and the situation it is used in.

Think about what you already know about the topic and where you might have seen a similar situation. Look for any key words and think about what they mean. Scribble what you know, particularly any important ideas that are related to this topic.

Now think about the information you need for this specific question. Some of what you have written may not be helpful.

Next, apply your science ideas to the new situation. Make a list of useful scientific words you could include in your answer.

Now write your full answer. Remember to write in full sentences and use scientific words.

Do a final check to make sure you have fully answered the question. Is everything the question asked about in your answer?

You can use the writing frame with any exam-style questions you want to practise that use the same command word. Below are some questions you can try to get you started.

Session 1

The diagram shows a palisade cell from the leaf of a plant.

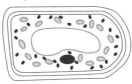

Palisade cells have lots of chloroplasts.

Suggest how the palisade cells are adapted to their function. **[3 marks]**

Session 8

The camel in the diagram is adapted to live in a hot, dry environment.

a) i) Suggest **one** biotic and **two** abiotic factors of the camel's environment. **[3 marks]**

   ii) Suggest how the camel is adapted to its environment. **[2 marks]**

b) Some plants are also adapted to live in hot, dry places.

   Suggest **two** adaptations of a desert plant that would help it survive. **[2 marks]**

Session 11

Suggest the most appropriate field investigation for each of these situations:

a) estimating how many snails are in a pond **[1 mark]**

b) calculating the percentage coverage of weeds in a football field **[1 mark]**

c) measuring how the number of crabs changes from the waterline of a beach to the land. **[1 mark]**

Session 18

Crude oil is a mixture of hydrocarbons of different lengths.
Fractional distillation is used to separate the different parts of the mixture.

The diagram shows a fractional distillation column.

a) Explain why longer hydrocarbon chains have higher boiling points. **[1 mark]**

b) Describe how the process of fractional distillation is used to separate the mixture of hydrocarbon chains. **[1 mark]**

c) There is not as much demand for very long hydrocarbon chains as there is for shorter hydrocarbon chains that can be used as fuels and to make polymers. Name the process used to make more short chain hydrocarbons from long hydrocarbon molecules. **[1 mark]**

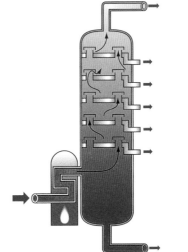

Session 21

Magnesium ribbon burns in air when it is heated in a hot, blue Bunsen flame.

a) Suggest why the magnesium only burns when it is heated to a high temperature in the Bunsen flame. **[1 mark]**

b) The magnesium burns faster if it is heated in pure oxygen. Use what you know about chemical reactions to suggest why magnesium burns faster when there is more oxygen available. **[2 marks]**

Session 21

A chemical reaction is the rearranging of atoms in the reactants to make new products with a different combination of atoms.

Suggest **three** ways to make chemical reactions faster. **[3 marks]**

Glossary

abiotic factor   physical or non-living conditions that affect the population in an ecosystem, such as light, temperature, soil pH

absorption   the process by which soluble products of digestion move into the blood from the small intestine

acceleration   the rate at which an object speeds up; calculated from the change in velocity divided by time; symbol a, unit metres per second squared, m/s^2

acids   compounds which produce hydrogen ions when dissolved in water and have a pH of less than 7

activation energy   the minimum energy needed for a chemical reaction to happen when particles collide

active site   the place on an enzyme where the substrate molecule binds

active transport   a process that uses energy to transport substances through cell membranes against a concentration gradient

adaptation   features that organisms have to help them survive in their environment

adult stem cells   rare, unspecialised cells found in some tissues in adults that can multiply to become many cells of one type

aerobic   respiration that involves the use of oxygen respiration

air resistance   the force produced by the collision of molecules in air with a moving object; the force acts to oppose the direction of movement

alkalis   compounds which produce hydroxide ions when dissolved in water giving solutions that have a pH of greater than 7; bases that are soluble in water

alkanes   a family of hydrocarbons with the general formula C_nH_{2n+2}

alkenes   a family of hydrocarbons with the general formula C_nH_{2n}

alleles   different forms of a gene

alloy   a mixture of a metal with one or more other metals or non-metals that changes the properties of the metal

alpha particle   a radioactive particle that is a helium nucleus (i.e. a helium atom without the electrons); emitted by an atomic nucleus during radioactive decay; alpha particles have a positive charge

alternating current   (a.c.) an electric current that continually changes direction

alveolus   (plural: alveoli) an air sac where gaseous exchange happens in the lungs

amplitude   the maximum displacement of a wave or an oscillating object from its rest position

amylase   a digestive enzyme (carbohydrase) that breaks down starch

anaerobic respiration   respiration without using oxygen

anode   the positive electrode in electrolysis

anomaly   data point that does not fit with the pattern of results; anomalies or anomalous results should not be included in calculations of the mean (average)

antibiotic   a medicine that works inside the body to kill bacteria (e.g. penicillin)

antibody   the protein made in response to an antigen, which it neutralises, to prevent infection

antigen   a substance that causes the body's immune system to react, especially by producing antibodies

antitoxin   the chemical produced by white blood cells (lymphocytes) that neutralises toxins

aorta   the artery that carries oxygenated blood from the left ventricle to tissues around the body

asexual reproduction   reproduction involving only one parent that results in genetically identical offspring

atomic number   the number of protons in the nucleus of an atom

atrium   (plural: atria) upper chamber of the heart that receives blood from the body or lungs

attract   cause two things to come closer together, e.g. the force that arises between a positive electric charge and a negative electric charge

base   a substance (metal oxide, hydroxide or carbonate) that neutralises an acid and produces a salt; a base that is soluble in water is called an alkali

becquerel   the unit of activity for a radioactive isotope, symbol Bq

beta particle   a fast-moving electron that is emitted by an atomic nucleus in some types of radioactive decay

biodiversity   all the different plant and animal species living in an ecosystem

biotic factor   the conditions caused by living things that affect the organisms in an ecosystem, e.g. food availability and competition between species

boiling point   the temperature at which a liquid turns into a gas

bronchus   (plural: bronchi) branch of the trachea – one bronchus goes to each lung

capillary network   the net of capillaries surrounding alveoli that provide a good blood supply

cardiac muscle   a special type of muscle found in the heart

catalyst   a chemical that speeds up the rate of a reaction but is not itself used up in the reaction

cathode   the negative electrode in electrolysis

cell   (in physics) circuit component that stores chemical energy in the form of chemical energy; lots of cells may be joined together to form a battery; (in biology) a small part of a living thing

cell cycle   a series of three stages during which a living cell divides

central nervous system   (CNS) the brain and spinal cord

charge   a property of matter, charge exists in two forms, positive and negative, which attract each other

chlorophyll   a pigment found in plants which is used in photosynthesis (gives green plants their colour)

chloroplast   a cell structure found in plants that contains chlorophyll

chromatogram   the paper that shows the results of a chromatography experiment; the distance each substance (e.g. dye) has travelled can be used to identify it by calculating the R_f value

chromatography   a method for separating substances, used to identify compounds and check for purity

chromosomes   thread-like structures in the cell nucleus that are made of DNA

circulatory system   organs (the heart and blood vessels) that work together to transport blood around the body

combustion   the exothermic reaction of a substance with oxygen

competition   when different species in an ecosystem need the same resources

compound two or more elements which are chemically joined together, e.g. H_2O (water)

compression a region of a longitudinal wave (e.g. sound wave) where the particles are closer together

concentration gradient the difference in concentration between two areas

concentric circles circles inside one another with the same centre point but different radii

condense when a gas changes into a liquid

conservation of mass principle stating that atoms (i.e. mass) cannot be created or destroyed

contact force the force that acts at the point of contact between two objects (e.g. friction)

contraception the method of preventing pregnancy

control variable a factor in an experiment that is kept constant so that it does not affect the results

coordination centre a region of the body that receives and processes information from receptors, e.g. brain, pancreas

coronary artery the vessel that supplies the heart muscle with oxygen and glucose for respiration

coronary heart disease a condition caused by fatty deposits in the coronary artery that reduces blood flow so less oxygen and glucose reach the heart tissue

covalent bonding chemical bonds between atoms where a pair of electrons is shared

cracking the process of breaking down large hydrocarbons into smaller molecules

crude oil a fossil fuel made millions of years ago; crude oil can be separated by distillation to give useful substances such as fuels

crystallisation a method of separation used to produce pure substances

crystals solids in which the ions are arranged in a regular lattice structure

current the rate of flow of electric charge; symbol I, unit amps (A)

cystic fibrosis a disorder of the cell membranes that is caused by a recessive allele

deceleration negative acceleration, when an object slows down

decomposition the process by which organisms break down the bodies of dead animals and plants

deforestation the removal of large areas of trees to provide land for cattle or growing crops

delocalised electrons electrons which are free to move in a giant structure or in graphite (a giant molecule)

density the mass of a substance divided by its volume

dependent variable the quantity in an experiment that is measured

diffusion the spreading out of the particles of a liquid, or a gas, resulting in a net movement from an area of higher concentration to an area of lower concentration

digestive enzymes biological catalysts that convert food into small soluble molecules that can be absorbed into the bloodstream in the small intestine

discharged released

distance–time graph a graph with distance on the y-axis and time on the x-axis; the gradient of a distance–time graph is equal to the speed

distillation the process of evaporation followed by condensation, used to separate a mixture of liquids with different boiling points

DNA (deoxyribonucleic acid) found as chromosomes in the nucleus – its sequence determines how our bodies are made

dominant refers to an allele that is expressed when one or two copies are present; represented by a capital letter

dot and cross diagram a diagram that shows the number of electrons in the outer shell of atoms or ions

double helix the shape of DNA – two strands of nucleotides that wind around each other like a twisted ladder

ductile a property of metals meaning they can be pulled into a wire

ecosystem all the living organisms and the non-living parts in an area

effector a muscle or gland that brings about a response to a stimulus

efficiency a measure of useful output energy transfer compared with the total input energy transfer; may be expressed as a percentage or as a decimal

efficient without much waste

electromagnet a magnet formed by an electric current flowing through a solenoid (coil of wire) with an iron core; the magnetic field of an electromagnet can be switched on and off by switching the current on and off

electrostatic attraction the force of attraction between opposite charges, e.g. between Na^+ and Cl^- ions

element a substance made of only one type of atom

electrolysis the process of passing a current through an ionic compound or a solution of an ionic compound in which the compound is broken down

electrolyte a liquid or solution that conducts electricity and breaks down during electrolysis

electromagnetic spectrum electromagnetic waves arranged in order according to their wavelength and frequency

electronic structure the arrangement of electrons in the sequence that they occupy the shells, e.g. the 11 electrons of sodium are arranged 2,8,1

electrons small negatively charged particles in an atom; they exist outside the nucleus in shells or energy levels

embryonic stem cells unspecialised cells found in early embryos that can differentiate into almost any cell type

emit to give out; emission – a process by which energy or a particle leaves an atom during radioactive decay (e.g. in beta particle emission, a high-speed electron is given out)

endangered species a species that is under threat of no longer existing (i.e. of becoming extinct)

endocrine system a control system in the body that communicates using chemical messengers, or hormones, to produce slow but long-lasting responses

endothermic a chemical reaction that takes in energy from the surroundings so the temperature of the surroundings decreases

enzyme a biological catalyst that increases the speed of a chemical reaction

equilibrium the condition when the forwards and backwards reactions in a reversible reaction in a closed system are occurring at the same rate

eukaryote an organism in which the cells contain a true nucleus in the cytoplasm (e.g. plant and animal cells)

Glossary

evolution   the gradual change in organisms over millions of years caused by random mutations and natural selection

exchange surface   a specialised area with a large surface area to volume ratio for efficient diffusion

exothermic   a chemical reaction in which thermal energy (heat) is given out to the surroundings

extension   the increase in length of an object when a force is applied to it

fermentation   anaerobic respiration in yeast cells; produces ethanol and carbon dioxide

fertilisation   the process by which a sperm or pollen enters an egg cell to form a fertilised egg

filtration   the process of using a porous material (or filter) to remove solids from water or solutions

flaccid   floppy; when there is not enough water in the vacuole of plant cells they shrink

follicle stimulating hormone   (FSH) a reproductive hormone that causes eggs to mature in the ovaries

food chain   a sequence of relationships in a community that starts with a producer (e.g. a green plant or alga)

foreign   describes a gene or antigen from a different species

fossil   the remains of organisms that lived millions of years ago

fractional distillation   a process in which a mixture of liquids is vaporised and compounds with different boiling points condense at different temperatures and can be collected separately (used to separate crude oil into fractions)

frequency   the number of waves passing a set point in one second

friction   the force acting at points of contact between objects moving over each other, to resist the movement

function   job

gametes   the male and female sex cells (sperm and egg cells in animals; pollen and egg cells in flowering plants)

gamma rays   ionising electromagnetic radiation with the shortest wavelengths in the electromagnetic spectrum

gas exchange   process in which one gas is taken in (e.g. carbon dioxide in leaves) in exchange for another given out (e.g. oxygen)

gene   a section of DNA that contains the instructions for a particular characteristic

genetic engineering   the transfer of specific genes from one organism to another

genome   the entire genetic material of an organism

genotype   the combination of the alleles present for a particular gene

giant covalent molecule   a large regular arrangement of atoms in a substance, held together by covalent bonds

giant crystal lattice   the regular three-dimensional arrangement of ions in an ionic compound; also called a giant ionic structure

global warming   the increase in the Earth's temperature due to increases in carbon dioxide levels in the atmosphere

glycogen   the main form in which glucose is stored in animals

GM crops   (genetically modified crops) varieties of crops that contain genes from another plant or organism

gradient   the steepness of a line on a graph

graphite   a substance made of layers of carbon atoms

gravitational field strength   a quantity that measures the 'pull' of the force of gravity on each kilogram of mass; symbol g, unit newtons per kilogram (N/kg)

half-life   the average time it takes for half the nuclei in a radioactive element to decay, or the time it takes for the count rate to halve

heterozygous   the presence of two different alleles for a characteristic, e.g. someone with blond hair may also carry the allele for red hair

homeostasis   the regulation of internal body conditions, such as temperature

homozygous   the presence of two alleles that are the same for a characteristic, e.g. a blue-eyed person will have two 'blue' alleles for eye colour

horizontal   a flat straight line running across the page parallel with the x-axis of a graph

hormone   a chemical messenger that acts on target organs

hydrocarbon   a compound containing hydrogen and carbon only

hydroxide ions   ions with a negative charge (OH^-); released by alkalis in aqueous solutions

immune response   how your body reacts to and destroys a pathogen

immune system   the body's defence against pathogens

independent variable   the quantity in an experiment that is changed

induced magnet   a material that is magnetic only when it is placed in the magnetic field of another magnet (e.g. the iron core within a solenoid)

infrared   electromagnetic radiation emitted by heated objects

insoluble   a substance that does not dissolve in a solvent such as water

interbreeding   organisms of the same species can interbreed to give fertile offspring

interdependence   when organisms depend on each other for survival

intermolecular force   a force between molecules in a substance

internal energy   the sum of the kinetic energy and potential energy of all the particles in a system

ionic bond   chemical bond between ions of opposite charge

ionic compound   a giant structure of ions held together by strong electrostatic forces of attraction between oppositely charged ions

isotopes   atoms with the same number of protons but different numbers of neutrons

kinetic energy   the energy that moving objects have

lipase   an enzyme that digests fats into fatty acids and glycerol

lock and key theory   a theory to explain how enzymes work; the substrate is the 'key' and the active site is the 'lock'

longitudinal wave   wave in which the vibrations of the particles are parallel to the direction of energy transfer (e.g. sound waves)

luteinising hormone   (LH) a menstrual cycle hormone that stimulates an egg to be released from an ovary

lymphocytes   white blood cells that produce antibodies and antitoxins to destroy pathogens

magnet   a material that produces its own magnetic field and so will attract other magnetic materials such as iron

magnetic field   the area around a magnet or current-carrying wire, where there is a force on magnetic materials

magnetic field lines a visual tool used to represent magnetic fields

magnitude the size of a quantity

malleable can be shaped or flattened

mass number the total number of protons and neutrons in an atom

matter any substance that has mass and takes up space

meiosis cell division that results in gametes being produced, with half the number of chromosomes as the parent cell

melting when a solid turns into a liquid

menstrual cycle the monthly cycle in females which is controlled by reproductive hormones

metallic bond the attraction between metal atoms due to delocalised electrons

microorganisms tiny organisms that can only be viewed with a microscope – also known as microbes

microvilli tiny projections found on villi in the small intestine; they increase the surface area for absorption of digested food molecules

microwaves electromagnetic radiation with a range of wavelengths longer than infrared but shorter than radio waves; used to cook food and for satellite and mobile phone communication

mitochondria (singular: mitochondrion) structures in a cell that carry out respiration

mitosis cell division that results in genetically identical diploid cells

mixture two or more elements or compounds not chemically combined together; mixtures can be separated by physical processes such as filtration or distillation

monomer a single repeating unit of a polymer; for example, nucleotide monomers are joined together in DNA (a polymer)

motor neurone a nerve cell that carries information from the central nervous system to the muscles

mutation a change in DNA that can result in a change in a characteristic

myelin a fatty insulating layer that surrounds neurones and speeds up nerve transmission

natural selection the process by which advantageous characteristics can be passed on and become more common in a population over many generations

neutralisation the reaction that takes place when an acid and a base react to produce a salt and water

neutron a particle with no charge found in the nucleus of an atom

neurone a nerve cell that is specialised to transmit electrical signals

Newton's First Law if the resultant force acting on an object is zero, a stationary object will remain stationary and a moving object will keep moving at a constant speed in a straight line

Newton's Second Law a resultant force on an object makes it accelerate in the same direction as the force; the acceleration is proportional to the magnitude of the force, and inversely proportional to the mass of the object; the equation for this is $F = ma$

Newton's Third Law when two objects interact, they exert equal and opposite forces of the same type on each other

noble gas a stable unreactive element found in Group 0 of the periodic table

non-contact force a force that acts at a distance between two objects that are not touching (e.g. a force due to an electric, gravitational or magnetic field)

non-renewable resource a resource that is used up at a faster rate than it can be replaced (e.g. fossil fuels)

nucleus (in biology) the part of the cell that contains the genetic material (DNA); (in chemistry and physics) the central part of an atom that contains protons and neutrons

oestrogen the main female reproductive hormone produced by the ovaries

ore a rock containing metals

organ a group of tissues that carries out a specific function

osmosis the diffusion of water molecules through a partially permeable membrane, from a dilute solution to a concentrated solution

palisade cells tightly packed cells found on the upper side of a leaf that carry out photosynthesis

pancreas the organ that controls blood sugar levels by releasing insulin

parallel circuit an electric circuit with two or more paths

partially permeable membrane a membrane that allows some small molecules to pass through

particle model a representation of the structure of solids, liquids and gases

pathogen a harmful microorganism that invades the body and causes infectious disease

peer review the process in which scientific experiments, writings and theories are checked by other scientists

penicillin an antibiotic, isolated from *Penicillum* mould, which was discovered by Alexander Fleming

periodic table a table of all the chemical elements arranged in order of their atomic numbers

peripheral relating to the edge of an area

permanent magnet a material that produces its own magnetic field

permeable allows small molecules to pass through

pH a measure of the number of hydrogen ions in a solution; acidic solutions have a pH of 0–6, neutral solutions have a pH of 7, alkaline solutions have a pH of 8–14

phagocyte a type of white blood cell that enters tissues, engulfs pathogens and then ingests them

phenotype the characteristic that is shown or expressed

photosynthesis the process carried out by green plants where sunlight, carbon dioxide and water are used to produce glucose and oxygen

poles the areas of a magnet where the magnetic field is strongest; magnetic poles always appear in pairs (one north, one south)

pollination the process of transferring pollen between one flower and another to enable fertilisation

polydactyly a condition caused by a dominant allele in which the sufferer has extra fingers or toes

poly(ethene) a polymer made of repeating units of ethene

polymer a very large molecule made up of repeating units (monomers) linked together

polymerisation the joining together of repeating units to form polymers

potential difference (p.d.) a measure of the energy transferred between two points in a circuit; also called voltage

Glossary

potential energy   the energy an object has because of its position or how its particles are arranged; e.g. the amount by which a material is stretched (elastic potential energy)

power   the rate at which energy is transferred or the rate at which work is done; an energy transfer of 1 J/s is equal to a power of 1 W (watt)

precipitation   the release of water droplets as rain, snow or sleet from clouds (in chemistry: the solid produced when two solutions react)

probability   the chance of an event happening; probability is given as a percentage, ratio or decimal

product   a substance made in a chemical reaction (shown on the right-hand side of the chemical equation)

progesterone   a reproductive hormone that causes the lining of the uterus to be maintained

prokaryote   a single-celled organism with DNA not enclosed in a nucleus

property   feature or characteristic

proteases   digestive enzymes that break down proteins into amino acids

protons   positively charged particles found in the nucleus of an atom

pulmonary artery   the vessel that carries deoxygenated blood from the right ventricle to the lungs

pulmonary vein   the vessel that carries oxygenated blood from the lungs to the left atrium

Punnett square   a grid used to find the possible outcomes of a genetic cross

quadrat   a square grid of known size (e.g. 0.5 m × 0.5 m) used to measure the distribution or number of species in an ecosystem

radiation   energy given out in the form of electromagnetic waves or moving particles; e.g. in radioactive beta decay a nucleus emits high-speed electrons, and the Sun radiates electromagnetic waves, including visible light

radioactive   describes a substance that exhibits radioactivity decay

radioactive decay   the process of an unstable atomic nucleus giving out radiation as it changes to become more stable

radioactivity   the process in which particles or energy are produced by the reactions of unstable atomic nuclei

radio waves   electromagnetic radiation with a range of wavelengths longer than microwaves; used for long-distance communication

random   no regular pattern

rarefaction   a region of a longitudinal wave (e.g. a sound wave) where the particles are further apart

reactant   a chemical that reacts with another chemical in a reaction (shown on the left-hand side of the chemical equation)

reaction profile   a diagram that shows the energy levels of the reactants and products and energy changes in a chemical reaction

receptors   cells in the body that detect changes in the environment

recessive   two copies of a recessive allele must be present for the characteristic to be expressed; represented by lowercase letters

reduction   when a reactant loses oxygen or gains electrons

reflex action   rapid automatic response to a stimulus

refraction   the bending of a wave as it travels from one medium to another at an angle

relative atomic mass   the sum of the mass of the protons and neutrons found in the nucleus of at atom

relay neurone   a neurone found in the spinal cord that transmits impulses from a sensory to a motor neurone

renewable resource   a resource that can be rapidly replaced

repel   to force or push away; a repulsive force acts between two objects and pushes them apart, e.g. the force between two positive electric charges

representative   data that is typical for the whole area being sampled

resistance   the opposition of an electrical component to the flow of current through it; symbol R, unit ohms (Ω)

resistor   an electric component that produces resistance to a current

respiration   the process used by all organisms to release energy from glucose

resultant force   the single force that would have the same effect on an object as all the forces that are acting on the object

reversible   describes a chemical reaction where the reactants form products that can react together to reform the reactants

ribosome   a structure in a cell where protein synthesis takes place

root hair cell   specialised cells in plant roots that are adapted for the uptake of water by osmosis and mineral ions by active transport

salt   a compound formed when an acid is neutralised by a metal or base (metal oxide, hydroxide or carbonate)

scalar quantity   a measurable quantity that has only a magnitude, not a direction (e.g. mass)

secondary sex characteristics   features that develop during puberty as a result of sex hormones being released, such as a deep voice and hair growth in boys, and breast development in girls

selective breeding   the process of breeding organisms with desired characteristics (also known as artificial selection)

sensory neurone   a nerve cell that carries information from receptors to the central nervous system

series circuit   an electric circuit in which all components are connected in a single loop

sex chromosomes   the pair of chromosomes that determines gender; XX in females, XY in males

solenoid   a coil of current-carrying wire that generates a magnetic field

soluble   a substance that can dissolve in a liquid, e.g. sugar is soluble in water

solvent   the liquid used to dissolve a solute to make a solution

speed   the distance travelled by an object per unit of time; unit metres per second (m/s)

specialised   when cells or tissues are adapted to carry out a specific function

species   a category of biological classification; members of the same species resemble one another, can breed among themselves, but cannot breed with members of another species

specific   relating to one thing and not others

stem cell   an undifferentiated cell of an organism which is capable of giving rise to many more cells of the same type, and from which certain other cells can arise from differentiation

stimulus a change detected by a receptor that causes a response by an effector

stopping distance the total distance a vehicle travels before coming to a complete stop; stopping distance = thinking distance + braking distance

sublimion the change of state of a substance from a solid directly to a gas; e.g. iodine

substrate a reactant that binds to the active site of an enzyme

tension the force that pulls or stretches

terminal velocity the constant velocity that occurs when the gravitational force acting downwards on a body falling through a fluid (e.g. air) is exactly balanced by the upwards force due to the resistance of the fluid

therapeutic cloning the process of creating stem cells with the same genes as the patient, through nuclear transfer

thermal energy energy that can be transferred as heat

tissue a group of cells that work together, with a particular function

trachea (windpipe) the tube through which air travels to the lungs

transpiration the movement of water up through a plant and its loss from the leaves

transverse wave a wave in which the vibrations of the particles of the medium are at right angles to the direction of energy transfer (e.g. water waves or electromagnetic waves)

turgid firm – plant cells which are full of water with their walls bowed out and pushing against neighbouring cells

Type 1 diabetes a condition where the pancreas cannot produce enough, or any, insulin

Type 2 diabetes a condition where the pancreas does not produce enough insulin and/or the body cells no longer respond to insulin

ultraviolet electromagnetic radiation with a range of wavelengths shorter than visible light but longer than X-rays; emitted by the Sun, for example

uncertainty a measure of the range about the mean; uncertainty is reduced when accuracy and precision are increased

undifferentiated describes a cell that can become many more cells of the same type, or a new type can arise by differentiation

uniform even, regular

universal indicator a substance that changes colour to show the pH of a solution

uterus (womb) the part of a woman's body where a fertilised egg may implant and develop into a baby

vaccine a medicine that contains an inactive or dead form of a pathogen that causes an immune response to prevent disease

vapour particles of a liquid suspended in air

variation the differences between individuals brought about by both genetic and environmental factors

vector quantity a measurable quantity that has both a magnitude and a direction (e.g. velocity)

velocity the speed of an object in a particular direction; symbol v, unit metres per second (m/s)

velocity–time graph a graph with velocity on the y-axis and time on the x-axis; the gradient of a velocity–time graph is equal to the acceleration

vena cava the vein that carries deoxygenated blood from the body to the right atrium

ventilate to move air into and out of the lungs

ventricles the lower chambers of the heart that pump blood around the body (left ventricle) or to the lungs (right ventricle)

villi structures on the inside surface (lumen) of the small intestine; they are covered in microvilli, which increase the surface area for absorption

virus a pathogen that reproduces inside cells causing cell damage

visible light electromagnetic radiation with a range of wavelengths shorter than infrared but longer than ultraviolet; detectable with the human eye

wavelength the distance between two wave peaks or the distance between identical points in adjacent cycles of a wave

weight the measure of the force of gravity on an object

white blood cell one of the body's defences against pathogens; white blood cells produce antibodies or antitoxins; they can also engulf pathogens by phagocytosis

wilt description of a plant that has drooping leaves due to lack of water

X-rays ionising electromagnetic radiation with a range of wavelengths shorter than ultraviolet and longer than gamma rays; used in X-ray photography to generate pictures of bones or teeth

yield the amount produced; (in chemistry) the amount of product produced during in a reaction, often expressed as the percentage yield

Solutions

1. Biological structures and their functions

Muscle cells and nerve cells are both involved in movement. Compare the structure and function of these cells.

Which structures do muscle cells and nerve cells have in common?

- Both cells have a cell membrane, nuclei and cytoplasm.
- Both cells have lots of mitochondria.

How is the shape of each cell different?

- Nerve cells are more branched at the ends than muscle cells.
- Nerve cells are much longer than muscle cells.

Which structures are different in muscle cells and nerve cells?

- Muscle cells have protein filaments.
- Nerve cells have myelin.
- Nerve cells have many connections at both ends.

For each of the differences, explain how that structure or shape helps the cell do its job.

- Muscle cells have protein filaments that slide over each other to move the muscle.
- Nerve cells are very long so they can transmit impulses over a long distance.
- Nerve cells have a myelin sheath which acts as an insulator and so makes the transmission of impulses faster.

Now bring together your answers: What do both cells have that they use to make movement happen? What does each cell have that is different that helps it to do its specific job?

- Both cells have mitochondria that release energy – the muscle cell uses the energy to move filaments over each other, and the nerve cell uses the energy to create an electrical impulse.
- Muscle cells are specialised because they have filaments that allow them to shorten, and nerve cells are very long and have a myelin outer layer so they can conduct electrical impulses over a distance quickly.

2. Moving molecules in living things

Plants need water for important reactions like photosynthesis.
Explain why plants wilt when they are not watered.

Which plant sub-cellular structures are involved in the structure of the plant?

Plant cell wall, vacuole.

Which plant organs does the water move through from the soil to the leaf? You could list these in order.

Roots → stem → leaves.

How does the water get into the plant? Name and define the process.

Through osmosis in the root hair cells. Osmosis is the diffusion of water from a dilute solution to a concentrated solution through a partially permeable membrane.

How does water move from the roots to the leaves?

Water is pulled up through the xylem vessels in the stem because of transpiration.

Plants lose water from their leaves. How does the water move out of the plant?

Water moves into the leaves by osmosis and is lost from the leaves through the stomata by evaporation from the air spaces.

What happens to the structures in plant cells when they lose too much water?

The vacuole shrinks away from the cell wall making the cell flaccid.

Now bring together what you have written to explain why plants wilt when they do not have enough water.

You will need to include:

- *what water is used for in plant cells*
- *how the water moves in and out of the cells*
- *what happens to the plant cells when the water has gone.*
- Water is needed for the structure of the plant, metabolic processes and photosynthesis.
- Water moves into and out of cells through osmosis because of a difference in solution concentrations. It moves through the plant due to transpiration.
- If plants cells lose too much water, they wilt. This is because the plant cells become flaccid as the vacuole pulls away from the cell wall and

they cannot replace the water lost in evaporation. Without watering they would eventually die.

3. Life systems

In order to move, animals need energy from respiration.
Compare how the oxygen and glucose needed for respiration enter the body and reach the cells.

For oxygen: Where does oxygen come from? How does it get in? List the parts of the respiratory system that the oxygen passes through. Name the organ where the oxygen meets the blood.

Oxygen comes from the air. It goes through the mouth or nose, then into the lungs. The oxygen travels to the lungs through the trachea, bronchi and into the alveoli. In the alveoli oxygen diffuses into the blood capillaries.

Glucose is found in food that needs to be digested before it can be absorbed. Food is broken down by enzymes. What do enzymes do? What do digestive enzymes do?

Enzymes are biological catalysts that speed up chemical reactions. Digestive enzymes help to break down large insoluble molecules into small soluble molecules.

The digestion of glucose happens in the digestive system. List the organs of the digestive system in order, from the mouth to the anus. Name the organ where the glucose meets the blood.

A – mouth, B – gall bladder, C – liver, D – large intestine, E – small intestine, F – stomach

Glucose meets the blood in the small intestine.

Oxygen and glucose both diffuse into the blood. What is diffusion?

Diffusion is the spreading out of the particles of any substance in solution, or particles of a gas, resulting in a net movement from an area of higher concentration to an area of lower concentration.

How are the organs where diffusion happens similar?

- They have a thin layer of cells.
- They have capillary networks.

Bring together your answers in a table: How are the journeys the same? How are they different? You can use tables in the exam for comparison questions.

	Oxygen	Glucose
How substance enters the body	Inhaled in air into the lungs Diffuses into the blood in the alveoli	Ingested in foods containing carbohydrates Moves into the blood in the small intestine
How substance enters cells	Diffuses from the blood into the cells	Diffuses from the blood into cells

4. Staying alive

Coronary heart disease is caused by lifestyle factors.
The graph shows the results of a study that looked at the effects of physical activity in men who had heart attacks.

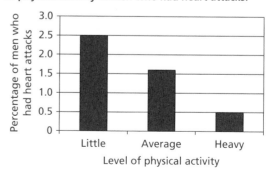

Suggest why people with coronary heart disease get out of breath easily. Use data from the graph to support your answer.

The heart is part of the circulatory system. What are the three types of blood vessel found in the circulatory system?

Arteries, veins and capillaries.

When you exercise your muscle cells need more energy from respiration. What is needed for respiration?

Oxygen and glucose.

Where do the substances needed for respiration come from?

Oxygen comes from the air through the lungs. Glucose comes from the diet through the small intestine.

How does the circulatory system move the substances to the muscles?

Oxygen diffuses from the alveoli into the blood capillaries surrounding them. The heart pumps the blood through the arteries and then the capillaries that surround the muscle.

At the muscle cells oxygen and glucose move from the blood into the muscle cells.

Coronary heart disease is caused by lifestyle factors. Which lifestyle factors can lead to coronary heart disease?

Lifestyle factors include poor diet, lack of exercise and smoking.

What does the data in the chart show you? Describe the results. Remember to include numbers from the graph.

The graph shows that the greater the amount of physical activity the lower the percentage of men who had heart attacks. Men who did little exercise had a 2.5% chance of having a heart attack, but men who did heavy exercise only had a 0.5% chance.

What happens to your breathing and heart rate when you exercise?

Both breathing and heart rate increase during exercise. Breathing rate increases the rate of gas exchange and heart rate increases the rate of blood reaching the cells.

Why does this happen?

Increased exercise leads to more muscle contraction which needs more energy from respiration. The increased oxygen needed for respiration comes from a faster ventilation rate. This also helps the carbon dioxide made by respiration leave the body. Respiration also needs glucose and the increased heart rate allows the blood carrying oxygen and glucose to reach cells faster.

Now bring together what you have written about how the heart works and what happens when you exercise to answer the question: Suggest why people with coronary heart disease get out of breath more easily.

With coronary heart disease the blood vessels around the heart become blocked. This means the heart cannot pump as well, so less oxygen and nutrients get to the cells, and more carbon dioxide builds up. To make up for this your breathing rate increases and you get out of breath as your body tries to get more oxygen into the body and more carbon dioxide out of the body.

5. Nervous control

Humans have reflexes to protect them from harm.

Show how a reflex arc makes you quickly move your hand away from something hot.

Which organs are in the central nervous system?

Brain and spinal cord.

Neurones or nerve cells are specialised to conduct nerve impulses. How are neurones specialised?

Neurones are very long which allows cells to connect the central nervous system to all parts of the body. There are fine branches which allow neurones to connect with other cells. Nerves are insulated by a myelin sheath so they can transmit information quickly.

What are the three types of neurone found in a reflex arc?

Sensory neurone, relay neurone, motor neurone.

Draw the reflex arc and list the steps. Make sure you include the names of each type of neurone.

See Key idea card B5.4.

Now bring together all the steps in order and link the fast reflex to the need to move quickly away from the heat.

When the hand touches something hot it needs to move quickly because...

The heat can burn and damage the cells.

The hand moves quickly because this is a reflex action. The steps of a reflex action are...

The stimulus is the hotplate. When the hand touches it the skin receptors detect the change in temperature and send an impulse to the spinal cord (central nervous system) through the sensory neurones. The spinal cord coordinates the response and sends an impulse through the motor neurones to the effectors, which are the muscles in the arm and hand. The response is to move the hand away.

What is another example of a reflex action?

Blinking when something comes close to our eyes.

6. Hormonal control

The graph shows how the amount of glucose in a healthy person's bloodstream changes after eating a meal.

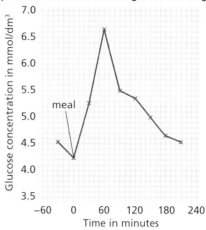

a. **Explain how your body controls blood sugar, by homeostasis, using hormones.**

b. **Diabetes is a condition in which blood sugar is uncontrolled.**

   Explain how the shape of the graph would look different in someone with diabetes.

For part (a): Where does the sugar in the blood come from?

Glucose (sugar) comes from carbohydrates in the diet.

For part (a): How is glucose (sugar) removed from the blood?

Glucose is stored as glycogen in the liver.

For part (a): How does the body know if there is too much glucose in the blood?

The pancreas detects sugar levels.

For part (a): Now put it all together to explain how your body controls blood sugar.

After a meal your blood glucose increases. The pancreas detects the increase in glucose and releases insulin. Insulin acts on your cells to take in glucose and store it as glycogen and use it in cell respiration. This leads to a drop in blood glucose.

For part (b): Look at the graph. Describe the shape of the graph in your own words.

The line goes down to start with then shoots up quickly after the meal. After the peak, it drops back down again.

For part (b): People who have a poor diet may develop Type 2 diabetes. What happens to blood sugar levels in someone with diabetes?

In Type 2 diabetes the pancreas will not make enough insulin and/or the cells do not respond to it. So blood sugar levels will remain high after someone with diabetes has eaten a meal.

For part (b): Draw a sketch on the graph in the question to show what would happen to blood sugar in someone with diabetes. Remember that your sketch is a good guess or estimate of how the line would look. How is your sketch different from the graph for the healthy person?

Your sketch should start slightly higher and then increase by the same amount up to 60 minutes but then should continue rising and stay higher for the rest of the graph.

7. Disease and immunity

Your immune system stops infections caused by pathogens (harmful microbes).

Explain how vaccines trigger this process and how this protects you.

What is the general term for an organism or virus that causes infectious diseases? What kind of organisms other than viruses can cause infectious diseases?

Infectious diseases are caused by pathogens. Pathogens are organisms like bacteria, fungi or protists that cause a disease.

What are the barriers that stop pathogens getting into the body?

Skin, nose hair and mucus, and stomach acid.

If pathogens do get in, how does the body recognise them?

Our body recognises any cell that is not a human cell. (This is because of the non-human proteins on their outer membranes.)

Solutions

Which cells are involved in fighting the pathogens?

White blood cells – phagocytes and lymphocytes.

What do these cells do?

Phagocytes ingest many types of pathogen. Lymphocytes make antibodies which stick pathogens together, and antitoxins which neutralise toxins.

Some of the white blood cells that have fought off an infection remain in the body. How does this help you if you are infected with the same pathogen again?

Once you have immunity, if you have an infection with the same pathogen your response will be faster and bigger.

Why does this only work for that specific pathogen?

Immunity comes from the lymphocytes. Lymphocytes are specific to a single type of pathogen.

Vaccines contain a tiny part of a pathogen, or a pathogen that has been damaged so it does not cause disease. How does a vaccine stop you getting ill?

When your body is infected with a tiny part of the pathogen your immune response is triggered. Specific lymphocytes for the pathogen are activated and they produce the specific antibodies needed to fight the pathogen. If you are infected with the real pathogen the secondary immune response is faster and bigger. The antibodies then stick to the pathogen and stop it from giving you the disease.

8. Life cycles

The materials that living things are made of have existed on Earth for millions of years. Describe how carbon moves through living things via respiration, photosynthesis and decomposition. Include the conditions needed for each process.

Plants are the first user of carbon, so we will start there. How do plants use carbon? What do they make it into? Write an equation for the process.

Plants take in carbon dioxide and convert it into glucose in photosynthesis.

$$\text{carbon dioxide} + \text{water} \xrightarrow{\text{light and chlorophyll}} \text{glucose} + \text{oxygen}$$

The glucose can be:

* used in cell respiration
* converted into other molecules to be used in growth and repair
* converted into cellulose for cell walls
* stored as starch.

Carbon can be passed on by the plant dying or being eaten. What happens to a plant if it dies? What are the conditions for decomposition?

When a plant dies it will decompose. Bacteria and other decomposers in the soil will recycle the plant back into nutrients which can be taken in by other organisms.

Carbon is passed along the food chain. Draw and label a food chain, naming each level. Use the labels: producer, primary consumer, secondary consumer, tertiary consumer.

For an example food chain, see key idea card B8.6.

Glucose contains carbon. Carbon in glucose is used in the process of respiration to release energy in plants and animals. Oxygen is needed, and carbon dioxide and water are produced. Show this as a word equation.

$$\text{oxygen} + \text{glucose} \rightarrow \text{carbon dioxide} + \text{water} + \text{energy}$$

Carbon can be released from fossil fuels when they are burned. Now sum up how carbon gets into organisms, and how it leaves. You could draw a labelled diagram.

For an example diagram of the carbon cycle, see key idea card B8.4.

9. Reproduction and inheritance

The organisms alive today are very different from the organisms that lived millions of years ago. Peppered moths are either white and speckled, or black. During the daytime peppered moths rest on branches of trees. In the 20th century the bark of these trees became covered in black soot.

Show how peppered moths got darker in colour through natural selection when the trees they lived on became sooty.

In the original population of moths, most moths were light coloured, and a few were dark. Describe why some organisms look different.

Variation in a population is caused by mutation, meiosis and sexual reproduction. Variation is why some organisms look different.

When the trees got sooty, more of the dark-coloured moths survived.

Why?

Dark-coloured moths were more likely to survive because it was easier for them to hide from predators on the dark sooty trees.

Camouflage like this is an adaptation to not being eaten. Animals and plants have many adaptations that help them survive. Write or draw four examples below.

There are many different adaptations. Some examples include:

* Polar bears have thick fur to insulate them from the cold.
* Woodpeckers have long tongues to take insects from crevices in trees.
* Cheetahs run fast to catch prey.
* Tigers have stripes to camouflage from prey.
* Flowers have bright colours to attract pollinators.

More of the next generation of moths were darker coloured. Why?

More dark moths were able to avoid being eaten by predators, so they were able to reproduce and pass on the dark colouring to more offspring.

Over time the whole population became darker coloured. Bring together the steps that led to this.

* Environmental change – trees became darker because of sooty pollution.
* Moths were a variety of different colours because of mutation, meiosis and sexual reproduction.
* Moths that were darker were better able to hide from predators.
* The darker moths survived and reproduced, having offspring which were also dark.
* Over the generations the population of moths grew darker.

10. Genetics and genetic engineering

Polydactyly is an inherited condition caused by a dominant allele.

A pregnant woman does not have polydactyly, but her husband Albert does.

Albert's mother does not have polydactyly.

Determine the possible genotypes and phenotypes for their child.

Complete the Punnett square shown to work out your answer.

Use D for the dominant allele and d for the recessive allele.

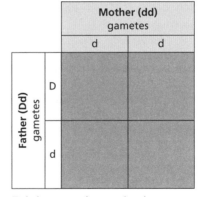

To help answer the question, let us recap some genetic terms. What is a dominant trait?

A dominant trait is one where only one allele of a gene needs to be present for the trait to be expressed.

What is the difference between a dominant and a recessive trait?

A dominant allele is always expressed if it is present in the genotype. Two copies of a recessive allele need to be present for the recessive trait to be expressed.

What is the difference between genotype and phenotype?

Genotype shows the alleles present for a gene – the genetic make-up. Phenotype is how the gene is expressed or shown in the individual.

In the question you are told to use D for the dominant allele and d for the recessive allele. Write down the genotypes of someone with polydactyly. There are two possible genotypes.

DD or Dd.

Now have a go at completing the Punnett square in the question. In each of the four empty squares, you need to write the letters of two alleles. One must come from the mother and one from the father.

See the Punnett square on page 181.

From the genotypes you have written for the offspring, identify which genotype is heterozygous, and which is homozygous.

The heterozygous genotype is: Dd

The homozygous genotype is: dd

Now identify the phenotype of each combination underneath the alleles in the Punnett square. In each of the four squares you should write 'polydactyly' or 'unaffected'.

		Mother (dd) gametes	
		d	d
Father (Dd) gametes	D	Dd polydactyly	Dd polydactyly
	d	dd unaffected	dd unaffected

Complete the sentences below to describe the probability of the child having polydactyly. You can write probability as a fraction, percentage or ratio.

The probability of the child having polydactyly is: 0.5, 1/2 or 50%

The probability of the child **not** having polydactyly is: 0.5, 1/2 or 50%

11. Biology practicals

A student carried out an experiment into the effect of light intensity on the rate of photosynthesis.

The student's results are shown in the table.

Distance of lamp from pondweed (cm)	Number of bubbles per minute		
	Test 1	Test 2	Test 3
10	99	121	124
15	41	50	54
20	20	30	32
25	11	17	16
30	8	13	14

Describe how the student could analyse their results.

What equipment would the student have used to measure the rate of photosynthesis?

Equipment – beaker, funnel, upturned measuring cylinder / test tube, aquatic plant.

To measure rate you need to measure how fast a product is made or how fast a reactant is used. Did the student measure a reactant or product in this investigation?

You would measure the rate of oxygen production (the product), by reading the volume collected in the measuring cylinder at regular time intervals.

How did they change the light intensity?

They changed the distance of the lamp from the plant.

Why does the rate of photosynthesis change when light intensity changes?

As light intensity increases more light can be captured by the chlorophyll in the leaves which increases the rate of photosynthesis. This means that more oxygen is produced as a (waste) product.

From the results shown you can see the student carried out three repeats (tests) of the experiment. What calculation could you do to get a representative result for the bubbles counted at each distance?

To get a representative result, calculate the mean (average).

Is the data repeatable or are there any anomalous results?

The results for test 2 and test 3 look repeatable, but test 1 results are anomalous – they do not agree. This could be because the student did not start counting the bubbles as soon as the timer was started. Or perhaps their method for counting the bubbles improved after the first test. The Test 1 results would need to be discounted from the calculated average or the test repeated a fourth time.

What kind of graph could they plot? What shape would the graph be?

Graph – a line graph with distance of lamp from pondweed (cm) along the *x*-axis and number of bubbles produced per minute on the *y*-axis. The sketch should show a downwards slope from left to right with a steeper slope to start with, levelling off towards 30 cm.

12. The particle model

The table shows the density of some different materials.

Substance	Density in kg/m³
Iron	8000
Gold	19 000
Water	1000
Air	1.3

a) **Explain why large molecules generally have higher melting points than smaller molecules.**

Use what you know about energy and changing state in your answer.

b) **Explain why different materials have different densities.**

Use data from the table and your knowledge of the particle theory to support your answer.

Draw the particles in solids, liquids and gases.

See key ideas card C12.2.

What is holding the particles in the patterns you have drawn?

Forces of attraction.

For part a): The particles you have drawn represent molecules. With bigger molecules, do they have more or less force holding them together?

They have more force holding them together.

For part a): When a substance is heated, the energy makes the substance warmer. When the substance changes state, the energy is used to break the forces that hold the particles of the substance together. What happens to the temperature when a substance changes state?

The temperature stays the same or remains constant.

For part a): Now bring together what you have written to explain why bigger molecules melt at higher temperatures.

Bigger molecules melt at higher temperature because it takes more internal energy to break the forces of attraction between the molecules. We can increase the thermal energy of the particles by heating. Therefore bigger molecules have a higher melting point.

For part b): Look at the data in the table. Write the state of each substance at room temperature (solid, liquid or gas).

Iron: solid. Gold: solid. Water: liquid. Air: gas.

For part b): Write down the formula that links density with mass and volume. How does particle size affect the density of a substance?

$$density = \frac{mass}{volume}$$

The larger the particle, the greater the density.

For part b): Now bring together what you have written to explain why different materials have different densities.

The density of a material depends on the size of the particles and how closely they are packed. So most liquids are less dense than solids and more dense than gases. The particles in a solid are packed closer together than in a liquid or gas.

Solutions

13. Atomic structure

The trends in physical and chemical properties in the periodic table depend on atomic structure and electronic structure.

Compare the reactivity of magnesium and calcium.

Find magnesium and calcium in the periodic table. Which group are they in? What does this mean?

They are both in Group 2. This means they both have two electrons in their outer shell.

How many protons, neutrons and electrons does an atom of magnesium have? How many protons, neutrons and electrons does an atom of calcium have? Use the periodic table to find out. Write your answers in the table.

	No. of protons	No. of neutrons	No. of electrons
Magnesium	12	12	12
Calcium	20	20	20

Draw the electronic structure of calcium and magnesium. Use crosses for electrons. Label the energy levels.

Magnesium:

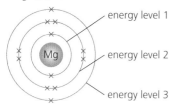

Calcium:

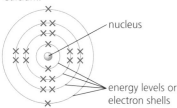

Magnesium is above calcium in the periodic table. Which one is more reactive? Why?

Calcium is more reactive. This is because, during chemical reactions, the outer electrons are 'transferred'. Down the group, the outer electrons are further away from the nucleus. Therefore, the electrostatic attraction with the nucleus is weaker and so it is easier to transfer electrons during a chemical reaction.

Bring together your previous answer to make a list of how magnesium and calcium are similar.

- Both are metals.
- Both are in Group 2 of the periodic table.
- Both have 2 electrons in their outer shell.

Bring together your previous answers to make a list of how magnesium and calcium are different.

- Calcium has more electron shells or energy levels.
- In calcium, the outer shell electrons are further away from the nucleus so the electrostatic attraction with the nucleus is weaker.
- Calcium is therefore more reactive than magnesium.

14. Calculating chemical change

Magnesium (a metal) reacts with oxygen (a gas) to form a compound (a white powder).

a) **Predict whether the solid product would be heavier, lighter or the same mass as the solid reactants.**

Use an equation to explain your prediction.

b) **Calculate how much magnesium oxide (MgO) would be made by the combustion of 5 g of magnesium (Mg).**

For part a): Decide what the reactants and products of the reaction are. Write a word equation.

Reactants: magnesium and oxygen; products: magnesium oxide.

magnesium + oxygen → magnesium oxide

For part a): Now write a formula equation.

The unbalanced equation is: $Mg + O_2 \rightarrow MgO$

For part a): Is the formula equation balanced? In other words, are there the same number of atoms of each element on both sides of the equation? If not, balance it.

$2Mg + O_2 \rightarrow 2MgO$

For part a): Now look at the equation again. Which reactants and which products are solids?

Reactant = Mg

Product = MgO

Now look at the balanced equation to compare how many atoms of the solid Mg (2) react to form how many molecules of the solid MgO (2). You can now predict whether the mass of product (2MgO) is heavier, lighter or the same mass as the magnesium reacted (2Mg).

The oxygen reactant is a gas, so the mass of reactant we need to compare is that of the solid magnesium only. The equation tells us that there is only one product, which is solid. Therefore, the mass of solid product (MgO) will be heavier than the mass of solid reactant (Mg).

For part b): The relative atomic mass of Mg is 24 and O is 16. Calculate the relative formula mass of MgO.

24 + 16 = 40

For part b): Using the equation $2Mg + O_2 \rightarrow 2MgO$, if we start with 24 g of Mg, how much MgO will be made?

40 g

For part b): When we start with 5 g of magnesium, the mass of magnesium oxide produced is scaled down by the same proportion. Work out the multiplier.

$\frac{5}{24} = 0.21$

For part b): Now bring together your previous answers to work out how much MgO is formed from 5 g of Mg.

40 × 0.21 = 8.4 g

15. Bonding

Element	Atomic number	Relative atomic mass	State at room temperature
Oxygen	8	16	Gas
Sodium	11	23	Solid
Chlorine	17	35.5	Gas
Sodium oxide			Solid

The table gives some information about the atoms of sodium, oxygen and chlorine.

a) **Suggest how sodium oxide crystals and molecules of chlorine gas are formed.**

b) **Explain why sodium oxide is a solid but chlorine is a gas.**

Use what you know about ionic bonding and covalent bonding in your answer.

For part a): Ionic bonds form when electrons are lost by one atom and given to another. What happens to the charge of an atom when an electron is lost? What happens to the charge of an atom when an electron is gained?

When an atom loses an electron, it is left with a positive (+) charge.

When an atom gains an electron, it gains a negative (−) charge.

For part a): How many electrons has each sodium atom lost?

One.

For part a): How many electrons has each oxygen atom gained?

Two.

For part a): How many sodium ions are needed for each oxygen atom? Write this number next to the sodium in the formula for sodium oxide. Make sure the N of sodium is capitalised and the a is lower-case. The O of oxygen is a capital.

Na_2O

For part a): Now write a sentence about what happens when an ionic bond is formed between sodium and oxygen.

During the formation of the ionic bond, each oxygen atom gains two electrons to form a 2− ion and each sodium atom loses an electron to form a + ion. The ionic bond is the electrostatic attraction between the ions.

For part a): Now bring together your ideas to write a sentence about what happens when chlorine atoms form covalent bonds to make a chlorine molecule.

The two chlorine atoms each share 1 electron from their outer shell.

For part b): Now bring together what you have written and try to explain why sodium oxide is a solid but chlorine is a gas.

Sodium oxide is a solid as the ions form a giant crystal lattice. It takes a lot of energy to break the forces of attraction between all the ions. Chlorine forms simple molecules made from two chlorine atoms. The forces between the molecules are very small.

16. Properties of materials

Graphite has two helpful properties: it conducts electricity and it is used as a solid lubricant.

a) **Explain why graphite has both properties.**

b) **Diamond is another form of carbon. Explain why diamond is used for drill tips whereas graphite is not.**

Use what you know about the how the carbon is bonded in each substance in your answer.

For part a): Carbon is in Group 4 of the periodic table. How many electrons does it have available for bonding in its outer shell?

Four electrons.

For part a): In graphite, each carbon atom only uses three of its available electrons for bonding, and has one free electron left over. The diagram below shows the structure of graphite. Add labels to the diagram to show the types of bond and the atoms present. What is the name given to the free electrons? See key ideas card C16.7.

The free electrons are called delocalised electrons.

For part a): Which of the features you have labelled makes graphite a solid lubricant? Explain this.

The weak intermolecular forces between layers of carbon atoms in graphite make graphite a solid lubricant.

For part a): Which of the features you have labelled makes graphite a conductor of electricity? Explain this.

The delocalised electrons make graphite able to conduct electricity.

For part b): What physical property does a material need if it is going to be used as a drill tip?

It must be very hard.

For part b): Suggest a reason why graphite is soft.

The weak intermolecular forces between the graphite layers mean they are easy to separate.

For part b): Diamond forms a giant molecular structure. Why is diamond harder than graphite?

Diamond uses all four of its available electrons for bonding. Every carbon atom is joined to four other carbon atoms by strong covalent bonds. This makes the structure very hard.

17. Elements, compounds and mixtures

Chromatography is a technique used to separate mixtures of different compounds or to test a compound for purity.

The diagram shows a chromatogram.

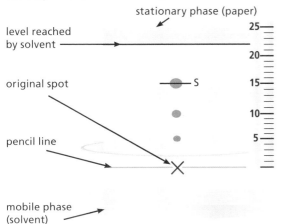

a) **Describe how chromatography is used to identify the dyes used in food colouring.**

b) **Calculate the R_f value of the spot labelled S on the diagram.**

For part a): Chromatography is used to separate mixtures. What is a mixture?

A mixture is a substance containing two or more different elements or compounds that are not chemically bonded together. They can be separated by physical processes.

For part a): Which physical property is used in chromatography to do the separation?

Solubility

For part a): What is meant by the terms mobile phase and stationary phase?

The mobile phase is the solvent that the mixture dissolves in.

The stationary phase is the chromatography paper that the solvent travels up.

For part a): Why is a pencil used to draw the starting line?

Pencil lead is insoluble (it does not dissolve in water).

For part a): Look at the diagram showing the chromatogram. How many dyes were found in the sample?

Three dyes (as there are three spots).

For part a): We can identify the dyes by calculating the R_f value. Write down the equation that links R_f values to the distance travelled by the spot.

$$R_f = \frac{\text{distance travelled by the spot}}{\text{distance travelled by the solvent}}$$

For part a): Now bring together these ideas to describe how chromatography is used to identify the dyes used in food colouring.

- Draw a pencil line on some chromatography paper.
- Spot the sample onto the line.
- Hang the paper in the chromatography tank.
- Let the solvent move up the paper.
- Leave the paper to dry.
- Calculate the R_f value.
- Compare the values to those of known dyes.

For part b): Calculate the R_f value of the spot labelled S on the diagram

$$R_f = \frac{15}{22} = 0.68$$

18. Earth chemistry

Earth's atmosphere was formed from chemical reactions and biological processes taking place over billions of years.

Crude oil is a fossil fuel formed millions of years ago.

Compare how photosynthesis and burning fossil fuels have changed the Earth's atmosphere.

What is photosynthesis? Describe photosynthesis in words or write the equation.

Plants make glucose from carbon dioxide and water by photosynthesis and produce oxygen gas as a by-product.

$$\text{carbon dioxide} + \text{water} \rightarrow \text{glucose} + \text{oxygen}$$

What is crude oil made from and where was it made?

Crude oil is made of the remains of dead organisms/sea animals. Crude oil was formed under the oceans.

Crude oil is a mixture of different substances. Describe how fractional distillation is used to separate them.

Crude oil is heated in a fractionating column until it forms a vapour. The vapour rises up the tower. The different hydrocarbons condense at the place in the column just below their boiling point. Short-chain hydrocarbons have lower boiling points than long-chain hydrocarbons so rise further up the tower. The different fractions are then collected.

What is combustion? Describe combustion in words or write the equation.

Combustion is when a fuel burns in oxygen to form carbon dioxide and water, giving out lots of energy.

$$\text{fuel} + \text{oxygen} \rightarrow \text{carbon dioxide} + \text{water}$$

Now think about the development of the Earth's atmosphere. When did photosynthesis start happening? When did combustion start happening?

Photosynthesis started when plants such as algae started to form billions of years ago when the atmosphere had lots of carbon dioxide. Combustion started to happen when humans began to burn fossil fuels a few hundreds of years ago.

Now bring together your answers to compare the two processes. You might like to think about what each process added or removed from the atmosphere, and when each one happened.

Solutions

Photosynthesis takes carbon dioxide out of the atmosphere and puts oxygen in whereas combustion uses oxygen and produces carbon dioxide. The early atmosphere had lots of carbon dioxide which gradually got replaced with oxygen as more plants started to grow. In more recent times the levels of carbon dioxide have started to increase as more fossil fuels have been burned.

19. Electrolysis

a) **Describe how pure aluminium is extracted from aluminium ore.**

b) **Explain why aluminium is so expensive.**

For part a), Where is aluminium found naturally?

Aluminium is found the ground as aluminium ore or aluminium oxide.

For part a): Why can aluminium not be extracted using carbon like iron can?

Aluminium is too reactive.

For part a): What are the names of the two electrodes used in electrolysis?

Positive electrode = anode; negative electrode = cathode.

For part a): Add labels to the diagram to show what happens at each electrode in electrolysis.

See key ideas card C19.5.

For part a): Describe or draw how electrolysis is used to extract aluminium. Add as many labels as you can.

See key ideas card C19.6 (aluminium extraction).

- Aluminium oxide is mixed with cryolite and heated until it melts.
- Cryolite lowers the melting point.
- An electric current is then passed through the liquid.
- Al^{3+} ions are attracted towards the cathode where the molten aluminium forms, which is then tapped off.
- The oxide ions are attracted to the anode where oxygen gas is formed.

For part b): Give two reasons why a lot of energy is used during the extraction of aluminium.

1. To melt the aluminium oxide.
2. Electricity is needed for electrolysis itself.

For part b): Why do the anodes need to be replaced regularly?

As soon as the oxygen is formed it reacts with the carbon to form carbon dioxide removing the carbon from the anode.

For part b): Now bring it together: Why is aluminium so expensive?

It takes a lot of energy to produce aluminium. Firstly to melt the aluminium oxide and then for the electrolysis itself. The carbon anodes regularly need to be replaced because as soon as the oxygen is formed it reacts with the carbon to form carbon dioxide; so the carbon anodes need to be replaced, which adds to the expense.

20. Acids and alkalis

Acids and alkalis are often used in homes in cooking, cleaning and medicines.

Sodium bicarbonate reacts with acids, neutralising them and producing a salt.

Explain why sodium bicarbonate is used in indigestion tablets.

Determine the reactants and products.

What is an acid?

A substance that produces H^+ ions in water; or, an aqueous solution of pH less than 7.

Why does extra stomach acid need treating?

Excess stomach acid may cause pain such as heartburn.

The acid in the stomach can be neutralised by sodium bicarbonate. Write a word equation for the reaction.

acid + sodium bicarbonate → a salt + water + carbon dioxide

The formula for sodium bicarbonate is $NaHCO_3$. Hydrochloric acid is HCl. Write down the formula of the three products. Write how many H atoms there are in a molecule of water. How many C atoms are there in carbon dioxide?

$NaCl$, CO_2, H_2O. There are 2 hydrogen atoms in water and 1 carbon atom in carbon dioxide.

Which product is a gas? What test would you carry out for the gas produced? Describe the test.

Carbon dioxide. Bubble the gas into limewater and if it goes cloudy then carbon dioxide is present.

The salt produced is called sodium chloride. To practise naming other salts, what is the name of the salt made when hydrochloric acid reacts with magnesium carbonate? Write a word equation for the reaction.

Magnesium chloride.

hydrochloric acid + magnesium carbonate → magnesium chloride + water + carbon dioxide

Naming salt practice: What is the name of the salt produced when sulfuric acid reacts with sodium hydroxide? Write a word equation for the reaction.

Sodium sulfate.

sulfuric acid + sodium hydroxide → sodium sulfate + water

Now, bring these ideas together to explain why sodium carbonate is used in indigestion tablets and give the reactants and products.

The acid in the stomach can be neutralised by sodium bicarbonate.

acid + sodium bicarbonate → a salt + water + carbon dioxide

Reactants: stomach acid, sodium bicarbonate

Products: a salt, water, carbon dioxide

21. Rates of reaction

Hydrochloric acid reacts with magnesium to produce magnesium chloride and hydrogen gas.

The graph shows how the volume of hydrogen collected varies with time at two different temperatures of hydrochloric acid.

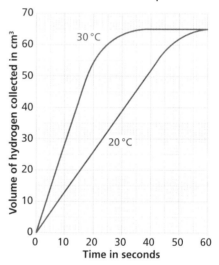

a) **Use collision theory to explain why the reaction is faster when the acid is more concentrated.**

b) **Suggest another way of increasing the rate of reaction.**

For part a): What happens to particles in a reaction?

The reactant particles must have a successful collision. This means they collide in the right direction and have enough energy to start the reaction.

For part a): What is the rate of a reaction?

It is the rate at which reactants are used up, or the rate at which products are made. It depends on the number of successful collisions that occur between molecules of the reactants per second.

For part a): What happens to the particles in a liquid when you heat it?

The kinetic energy of the particles increases.

For part a): What does this do to the rate of a reaction? Use the graph to support your answer.

The time taken for X amount of hydrogen to be produced decreases as the temperature is increased from 20 to 30 °C. For example, to produce 30 cm³ of hydrogen gas: at 20 °C it takes 24 seconds and at 30 °C it takes 12 seconds. So the rate of reaction increases.

For part a): Bring all this together to explain why the rate of a reaction increases when the reactants are heated.

As the reactants are heated, the kinetic energy of the particles increases and they move more quickly. This means that they will collide with other particles more often which will lead to in an increase in the number of successful collisions occurring each second, and so the rate of reaction increases.

For part b): Now list other ways a reaction can be speeded up.

- Increasing the concentration or pressure
- Increasing the surface area of the reactants.
- Using a catalyst.

22. Chemistry practicals

Acids react with some metals to give a salt and hydrogen. Thermal energy is given out during the reaction.

Design an investigation to show how changing the concentration of hydrochloric acid affects the temperature change seen as it reacts with zinc metal.

What equipment would you use for this procedure? You can show this using a diagram. List the range of concentrations you would use. How will you measure the temperature of the reaction? When will you measure this?

Equipment: thermometer, balance, polystyrene cup and lid, $250\,cm^3$ beaker, $50\,cm^3$ and $10\,cm^3$ measuring cylinders, hydrochloric acid ($1\,mol/dm^3$) and distilled water.

Use the thermometer to measure the temperature of the acid at the start and end of the reaction.

List the steps of the practical procedure.

- Cut a 5 cm length of magnesium metal and clean with sand paper until shiny.
- Measure out $30\,cm^3$ of acid and put it in the polystyrene cup. Stand the cup inside the beaker.
- Measure the temperature of the acid and record it in a table.
- Add the magnesium to the cup; put on the lid and gently stir with the thermometer through the hole.
- Watch the temperature rise. When the temperature stops rising record the temperature in the table.
- Repeat the experiment 5 more times but each time dilute the acid using distilled water. For example, in the second test use $25\,cm^3$ of acid and $5\,cm^3$ of water, etc.
- Finally, repeat each experiment and then work out the mean temperature change for each concentration of acid.

What are the hazards? What precautions would you take to avoid harm?

Dilute hydrochloric acid is an irritant to eyes and skin – wear safety glasses to stop splashes in the eyes; wash hands with water if it gets on skin.

Magnesium is flammable – make sure there are no naked flames around. Polystyrene cups can be knocked over easily – put the cup in a beaker to make it more stable.

What would you expect to happen?

As the concentration of the acid increases so does the temperature change of the solution. The temperature also goes up more quickly.

Why would you expect this to happen?

The reaction is exothermic which means energy is given out as a result of successful collisions between the reactant particles. At lower concentrations all the acid will be used up and the magnesium will be in excess. This means there will be fewer successful collisions taking place. At higher concentrations there will be more successful collisions taking place and more energy is given out. The temperature will go up more quickly at the start because there will be more successful collisions per second in a more concentrated acid.

23. Forces

Forces can affect the motion of objects.

The badminton shuttlecock shown below is falling downwards at constant speed.

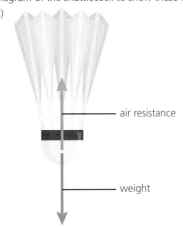

a) **State the law that gives the relationship between the two forces acting on the shuttlecock when it is moving at a constant speed.**

b) **Draw arrows on the diagram to show the two forces acting on the shuttlecock.**

c) **Describe what happens when the shuttlecock hits a stationary badminton racket facing upwards. You can use a diagram to help you explain.**

For part a): Which one of Newton's laws does this show? Decide which one deals with equal and opposite reactions.

Newton's Third Law.

For part b): If the shuttlecock is moving at a steady speed then the forces must be balanced. What are the two main forces? Add arrows to the diagram of the shuttlecock to show these forces.

a)

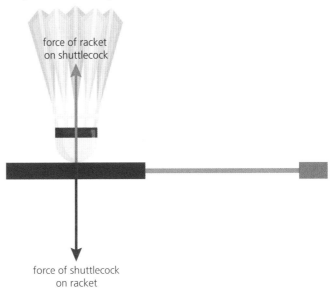

air resistance

weight

For part c): Try drawing the shuttlecock as it makes contact with the racket. Now add arrows to your diagram to show the two forces. How will you use the size of your arrows to show the size of the forces?

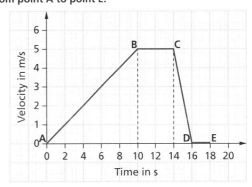

force of racket
on shuttlecock

force of shuttlecock
on racket

For part c): Describe what happens to the shuttlecock as it makes contact with the racket.

The shuttlecock exerts a downwards force on the racket. The racket exerts an equal and opposite force on the shuttlecock acting upwards. The shuttlecock rebounds/moves off the racket upwards.

24. Motion

The graph shows a velocity–time graph for a girl riding a bicycle from point A to point E.

a) **Describe, in words, the girl's journey from A to E.**

b) **i) Determine the change in velocity from A to B.**

  ii) Calculate the acceleration of the girl from A to B.

Solutions

c) Calculate the distance that the girl travels between B and C. Use the equation:

$$\text{distance (m) = velocity (m/s) × time (s)}$$

To answer part a), break the graph down into steps. In the space below, write what is happening between the labelled points on the graph. The question asks you to use words only so there is no need to write any numbers here.

The girl's journey from A to E is summarised as:

- A to B: moving with constant acceleration
- B to C: moving at constant velocity
- C to D: moving with a constant deceleration (value higher than the initial acceleration)
- D to E: stationary.

For part b) i) you first need to read values from the graph. Look at the y-axis value after 10 seconds. How much has this value changed from 0 seconds?

change of velocity = 5 m/s – 0 m/s = 5 m/s

For part b) ii), you need to calculate the acceleration from A to B. Put the values for change in velocity and time into the equation:
acceleration (m/s²) = change in velocity (m/s) / time taken (s). Remember to write the units.

$$\text{acceleration} = \frac{\text{change in velocity}}{\text{time taken}}$$

$$= \frac{5\,\text{m/s}}{10\,\text{s}} = 0.5\,\text{m/s}^2$$

For part c), what is the constant velocity between B and C? Then calculate the time between B (10 s) and C (14 s). Put these two values into the equation given:

$$\text{distance (m) = velocity (m/s) × time (s)}$$

What will the units be?

Between B (10 s) and C (14 s) the girl travels for 4 s.

$$\text{distance} = \text{velocity} \times \text{time}$$

$$= 5\,\text{m/s} \times 4\,\text{s} = 20\,\text{m}$$

25. Waves

The diagram shows a transverse water wave. The frequency of the wave is 3 Hz and the wavelength of the wave is 2 m.

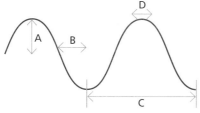

a) Identify the label corresponding to the:
   i) amplitude
   ii) wavelength.

b) Calculate the speed of the wave using the wave equation.

c) The table below shows the parts of the electromagnetic spectrum. Some of the parts are missing.

radio waves		infrared		ultraviolet		gamma rays

   i) Write the missing words in the table.
   ii) Draw an arrow under the table to show in which direction wavelength increases.
   iii) Explain why ultraviolet rays are dangerous to humans.

For part a): What are the characteristics of a transverse (e.g. water) wave? Where do we measure amplitude and wavelength from?

i) amplitude = A

ii) wavelength = C

For part b): Gather all the numerical information from the question that you need to use the wave equation:

$$\text{wave speed (m/s) = frequency (Hz) × wavelength (m)}$$

Then substitute the numbers into the equation and use a calculator to work out the answer. Repeat this on your calculator to check that you have entered the numbers correctly. What are the units of wave speed in this question?

$$\text{wave speed} = \text{frequency} \times \text{wavelength}$$

$$= 3\,\text{Hz} \times 2\,\text{m} = 6\,\text{m/s}$$

For part c) i), ii): List all of the parts of the electromagnetic spectrum in the correct order. Try to remember which have the longest wavelength, and so which way the arrow should point.

i)

radio waves	**microwaves**	infrared	**visible light**	ultraviolet	**X-rays**	gamma rays

ii) A horizontal arrow pointing to the left should be drawn under the table.

*For part c) iii): You need to **explain** in this question – this means you need to state the reasons for something. First, state the property of ultraviolet rays that makes them dangerous. Next, give the reasons why this makes them dangerous.*

Ultraviolet rays are dangerous to humans because they have a high energy and can harm or kill cells.

26. Energy

A man, with a mass of 75 kg, climbs a flight of steps 1.5 m high. At the top of the steps he stops, before jumping off the top step and landing back down on the ground.

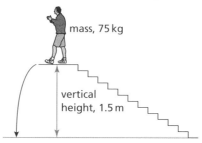

a) Describe the changes in energy as he climbs the steps, stops, and then jumps down.

b) Calculate the gravitational potential energy of the man at the top of the steps.
   The gravitational field strength is 10 N/kg.

c) The velocity of the man, just before he hits the floor is 5.2 m/s. Calculate the kinetic energy of the man just before he hits the floor.

d) Use your answers to parts b) and c) to calculate the energy wasted as sound energy and thermal energy as the man falls.

For part a), you need to describe how the energy transfers from kinetic energy into gravitational potential energy and back again. How is energy wasted?

Kinetic energy of the man moving up the steps is transferred to gravitational potential energy at the top of the steps, with some wasted as sound and thermal energy. This gravitational potential energy is then converted back into kinetic energy as the man falls, with some energy wasted as sound and thermal energy.

For part b), you need to recall the equation for gravitational potential energy:

$$\text{gravitational potential energy} = \text{mass} \times \text{gravitational field strength} \times \text{height}$$

Underline all the values you have been given in the question and/or on the diagram. Now insert them into the equation. Can you remember the units? Use a calculator to work out the answer. Remember to check your answer by repeating the calculation on your calculator.

$$\text{gravitational potential energy} = \text{mass} \times \text{gravitational field strength} \times \text{height}$$

$$= 75\,\text{kg} \times 10\,\text{N/kg} \times 1.5\,\text{m}$$

$$= 1125\,\text{J}$$

For part c), you need to recall the equation for kinetic energy:

$$\text{kinetic energy} = 0.5 \times \text{mass} \times \text{velocity}^2$$

Underline all the values you have been given in the question and/or on the diagram. Now insert them into the equation. Can you remember the units? Use a calculator to work out the answer. Remember to check this by repeating the calculation on your calculator.

$$\text{kinetic energy} = 0.5 \times \text{mass} \times \text{velocity}^2$$

$$= 0.5 \times 75\,\text{kg} \times (5.2\,\text{m/s})^2$$

$$= 1014\,\text{J}$$

For part d), remember that the law of conservation of energy states that energy does not just disappear. Therefore, the gravitational potential energy at the top of the steps is transferred into kinetic energy plus wasted energy.

energy wasted = gravitational potential energy – kinetic energy

Use the values from parts b) and c) in your calculation. Remember to write the units. Now use a calculator to work out the answer. Remember to check this by repeating the calculation on your calculator.

All the gravitational potential energy at the top of the steps will be transferred into kinetic energy plus the wasted energy, so:

energy wasted = gravitational potential energy – kinetic energy

$$= 1125\,J – 1014\,J$$
$$= 111\,J$$

27. Electricity

A student performs an experiment to measure the potential difference across, and current through, a fixed resistor. She collects the following results:

Potential difference in V	0	2	4	6	8	10
Current in A	0.0	0.1	0.2	0.5	0.4	0.5

a) Plot a graph of these results. Plot the potential difference on the x-axis and current on the y-axis.

b) One of the results has been recorded incorrectly and is an anomaly. Identify the anomaly and draw a circle around this result on your graph.

c) Draw a line of best fit through the valid data points.

d) Suggest a correct value for the anomalous result.

e) The equation for resistance is:
$$resistance = \frac{potential\ difference}{current}$$
Calculate the resistance of the fixed resistor.

Use the result for a potential difference of 8 V and the equation above.

For part a) you need to **plot** a graph. First, look at the data given. You are told which row needs to be plotted on the x-axis and which one on the y-axis. Does the graph start at (0, 0)? What are the maximum x-axis and y-axis values? Draw a scale that fits both using a pencil and a ruler (in case you make a mistake). Plot the data points accurately using the scales. Remember to label your axes using the table row headings.

For part b), can you see an **anomaly** (a data point that does not fit the pattern)? Circle it on your graph.

For part c), think about lines of best fit. Is this one a straight line, or a curve? Does it go through (0, 0)? Do you include the anomaly?

a) to c)

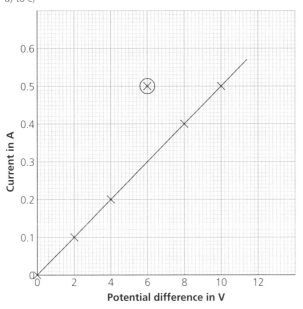

For part d), now use the line of best fit that you have drawn to identify the true value of the current at 6 V.

potential difference = 6 V; current = 0.3 A

For part e), read the value of the current off the graph corresponding to a potential difference of 8 V. Write down the data values that you have read off the graph, and the equation given. Insert the values into the correct place in the equation and use a calculator to calculate your answer. Repeat the calculation using your calculator to double check your answer. Write down your final answer and write down the unit of resistance. You can use the symbol or the word.

At a potential difference of 8 V, current = 0.4 A, so:

$$resistance = \frac{potential\ difference}{current}$$

$$resistance = \frac{8\,V}{0.4\,A} = 20\,\Omega\ (or\ 20\ ohms)$$

28. Electromagnetism

A child's toy magnet set, containing three solid metal rods, has become mixed up.

• One rod is a permanent magnet.

• The second rod is made from iron.

• The third rod is made from aluminium.

All the rods look the same.

Explain how a permanent bar magnet could be used to determine the nature of each rod.

What are the rules for the interaction of magnetic poles? How could these be used to identify the permanent magnet rod?

Like poles repel and unlike poles attract.

Bring each pole of the permanent bar magnet up to each rod in turn.

If the end of the rod is attracted to one pole of the permanent magnet, but it is repelled by the other pole, then the rod is a permanent magnet.

What is induced magnetism? What materials can be made into an induced magnet? Does it matter which pole of the permanent magnet is used to induce magnetism? How could this be used to identify the iron rod?

Induced magnetism occurs when a permanent magnet is brought up next to an object made from iron, steel, cobalt or nickel. The permanent magnet 'induces' (creates) a magnetic field in the object, making it behave like another permanent magnet. Both poles of the permanent magnet induce an opposite pole in the object next to it. It doesn't matter which way around the object is. If both ends of the rod are attracted to both poles of the permanent magnet, then the rod is made from iron, because iron can be made into an induced magnet with either of the ends of the permanent magnet.

What happens when a permanent magnet is brought up close to a material that cannot be magnetised? How could this be used to identify the aluminium rod?

If one end of the rod is not attracted to either pole of the permanent magnet, then the rod is made from aluminium, because aluminium cannot be magnetised.

Now bring together all this information to explain how to determine the nature of each rod.

The rod that is a permanent magnet is attracted to one pole of the permanent magnet but is repelled by the other pole.

The iron rod is attracted to both poles of the permanent magnet.

The aluminium rod is not attracted to either pole of the permanent magnet.

29. Atoms and radioactivity

The table below shows how the activity of a radioactive sample of lead-217 (atomic number 82) changes over time.

Time in s	0	10	20	30	40	50
Activity in Bq	120	85	60	42	30	21

a) Plot a graph of activity against time.

b) Draw a line of best fit through the points.

c) Use information from the graph to determine the half-life of lead-217.

d) Lead-217 decays by beta particle emission into an isotope of bismuth, Bi. Balance the nuclear decay equation for this decay:

$$^{217}_{82}Pb \rightarrow\ ^{0}_{-1}e +\ ^{...}_{...}Bi$$

Solutions

For part a): You will first need to determine the axes for the decay graph. Which quantity goes on which axis? What scale do you need – what are the lowest and the highest values that you need to plot on each axis? Do you need to start the graph axes at (0, 0)? Make sure that you plot the points accurately using a sharp pencil. Double-check them.

For part b): Think about the general shape plotted out by the points (is it a curve or a straight line)? Remember, a line of best fit does not need to go through all the points, but it does need to be a smooth line, not join-the-dots.

a) and b)

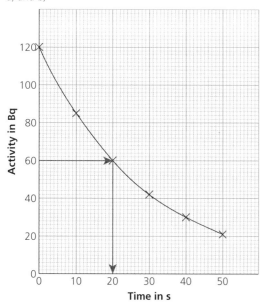

For part c): The half-life is the time taken for the initial activity to fall to half its value. What is the initial activity? What is half of this initial value? Draw a straight line across the graph from this value, so it touches the line of best fit. Then draw a line down to the time axis and read off the half-life. What are the units?

The horizontal arrow shows half the initial activity, this hits the line of best fit at a time of 20 s. This is the half-life.

For part d): You need to balance the nuclear decay equation. The top, mass numbers need to be the same on each side of the equation. What is the mass number of the bismuth isotope? The bottom, atomic numbers, also need to balance on either side of the equation. Remember the value for a beta particle is –1, so the atomic number of bismuth – 1 = atomic number of lead. What is the atomic number of the bismuth isotope?

$$^{217}_{82}\text{Pb} \rightarrow \,^{0}_{-1}\text{e} + \,^{217}_{83}\text{Bi}$$

The missing mass number = 217; the missing atomic number is 83.

30. Physics practicals

Design an investigation that shows the effect of increasing the mass of an object hung from a spring.

What equipment would you use for this procedure? You can show this using a diagram. List the range of masses you would use. How will you measure the extension of the spring?

Provided that the range of masses is low, springs obey the equation:

$$\text{force} = \text{spring constant} \times \text{extension}$$

The force (weight) of the hanging masses is given by:

$$\text{weight} = \text{mass} \times \text{gravitational field strength}$$

If the equation is correct, then adding more mass should increase the extension, and doubling the mass should double the extension.

You will need:

2 clamps

2 bosses

1 stand

1 30 cm ruler

1 pointer

1 spring

1 mass stack (6 ×100 g)

For a diagram, refer to key ideas card P30.6.

You can measure the extension directly off the ruler if you ensure that the zero of the ruler is in line with the pointer at the bottom of the unloaded spring. As masses are added, the spring stretches and the extension is measured off the ruler in line with the pointer.

List the steps of the practical procedure.

- Set up the apparatus as shown in the diagram.
- Add one mass to the mass hanger (100 g). Record the mass.
- Measure and record the extension of the spring using the pointer.
- Remove the mass stack from the spring. Check that the end of the spring has returned to zero on the ruler to ensure that no permanent stretching has occurred (and that the ruler has not moved).
- Repeat these steps for masses of 200 g to 600 g in 100 g steps.
- Remove the mass stack from the spring.
- Repeat these steps twice more so that you have three repeats for each added mass.
- Calculate the mean extension for each added mass.
- Plot a graph of extension (*y*-axis) against mass (*x*-axis).
- Draw a line of best fit through your points and state the relationship between the mass and the extension.

What are the hazards? What precautions would you take to avoid harm?

The main hazards in this experiment are:

- The mass stack falling off the spring and falling on a foot causing a bruise or a break. The risk from this can be minimised by performing the experiment in the middle of a laboratory bench.
- The stand falling over and hitting someone in the face causing a cut or a bruise or damaging an eye. The risk from this can be minimised by clamping the stand to the bench and wearing safety goggles.

What would you expect to happen?

If the spring obeys the equation:

$$\text{force} = \text{spring constant} \times \text{extension}$$

the graph of extension (*y*-axis) against mass (*x*-axis) should be a straight line through (0,0). Doubling the mass on the spring will double the measured extension.

Why would you expect this?

New, unstretched springs will obey the equation:

$$\text{force} = \text{spring constant} \times \text{extension}$$

provided that the range of 0–600 g is not exceeded.

Balance

Session 19
a) copper + oxygen → copper oxide

b) i) Cu(s) + O$_2$(g) → CuO(s)

ii) 2Cu + O$_2$ → 2CuO

Session 20
a) sodium carbonate + hydrochloric acid → sodium chloride + water + carbon dioxide

b) Na$_2$CO$_3$(s) + 2HCl(aq) → 2NaCl(aq) + H$_2$O(l) + CO$_2$(g)

Calculate

Session 1

a) i)

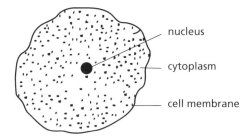

nucleus

cytoplasm

cell membrane

ii) magnification of the image = $\dfrac{\text{size of the image}}{\text{size of real object}}$

The cell in the diagram measures 50 mm across. In real life, it measures 40 µm.

Convert 50 mm into micrometres (or 40 µm to millimetres): 50 mm = 50 000 µm

The cell measures 40 µm, so the magnification of the image = 50 000 / 40 = ×1250

b) Take a thin layer of cells from a piece of onion.

Use forceps to place the layer of onion cells on a microscope slide.

Add a drop of water or iodine.

Lower a coverslip onto the cells using a mounted needle or forceps, taking care to not trap bubbles of air. Remove any excess liquid on the slide. View the slide under a light microscope.

Equipment list: Microscope slide, coverslip, forceps, iodine in dropping bottle, mounted needle.

Safety: Iodine will stain, so take care not to spill it on clothes or skin. Iodine can be an irritant, so wear safety goggles. The glass coverslip and slide can break easily – handle carefully.

Session 14
a) The reaction of copper carbonate with carbon is a chemical change because bonds are broken and made to form new compounds.

b) i) copper carbonate + carbon → copper + carbon dioxide

ii) 132 kg of carbon dioxide.

Total mass of reactants = 247 kg + 12 kg = 259 kg.

Mass is conserved, so 259 kg – 127 kg = 132 kg.

Session 24
a) In graph A the skateboarder starts 15 m away from the origin and is travelling at a constant speed. In graph B, the skateboarder is stationary, 20 m away from the origin, for 6 s. In graph C the skateboarder starts at the origin, moves away at a constant speed for 3 s to a distance of 25 m, and then moves back towards the origin at the same speed for 3 s.

b) i) speed = $\dfrac{\text{distance}}{\text{time}} = \dfrac{15\,\text{m}}{6\,\text{s}} = 2.5\,\text{m/s}$

ii) speed = $\dfrac{25\,\text{m}}{3\,\text{s}} = 8.3\,\text{m/s}$, for 3 s

c) Between 0 and 3 seconds in graph C.

d) Between 3 and 6 seconds in graph C.

Compare

Session 2
a) Mitochondrion.

b) glucose + oxygen → carbon dioxide + water

c) In both cases oxygen diffuses from the blood into the muscle cells. In both cases oxygen moves from an area of high concentration to an area of low concentration along the concentration gradient. With the person who is exercising, there will be less oxygen inside the muscle cells, as it is used up in respiration. Conclusion: in both people oxygen moves into the muscle cells by diffusion. Diffusion is faster in the person who is exercising as they use up the oxygen in the muscle cells faster, which increases the concentration gradient.

Session 4
a) A – aorta; B – left ventricle; C – right atrium; D – vena cava.

b) Similarities: both sides of the heart have a blood vessel (a vein) that delivers blood and another blood vessel (an artery) that takes blood away from the heart. Both sides of the heart have an atrium and a ventricle, with valves between them. Both sides of the heart have muscle to pump the blood out of the heart. The left side of the heart has more muscle as it has to pump blood around the body, whereas the right side has less muscle as it only has to pump blood to the lungs, which is not as far. Conclusion: both sides of the heart have the same structures, but the left side of the heart has more muscle to pump the blood further.

Sessions 5 and 6
Both actions result from a change being detected and result in a response by an effector. In both actions the change is detected by receptors. In both actions a control centre coordinates the response. This is the pancreas in the case of blood sugar and the spinal cord / CNS in the case of the reflex response. Reflex responses involve electrical signals via the nervous system; hormonal responses involve chemical signals via the blood system.

Session 10
Similarities: Both activities result in a change in characteristics. Both activities result in a change in genetic material. Differences: Selective breeding takes a long time – can be centuries. Genetic engineering can make changes more quickly. Genetic engineering is much more expensive. Genetic engineering is more predictable.

Session 26
Wind power is a renewable energy source; nuclear power is a non-renewable energy source. Wind power is unreliable; nuclear power is reliable. Huge numbers of wind turbines are required to produce the same electricity output as a nuclear power station. Wind turbines can only be placed in locations that are windy for most of the year; nuclear power stations can be located anywhere.

Session 29
a) Labels added for electron (outer circle), neutron and proton (central circles; either way round).

b) Similarities: protons and neutrons are roughly the same mass; protons and electrons both have a charge. Differences: protons have a positive charge whereas electrons have a negative charge. Neutrons do not have a charge whereas electrons and protons do. Protons and neutrons have a relative mass of 1 whereas electrons are tiny with a very small relative mass.

c) A helium atom consists of a nucleus of two protons and two neutrons surrounded by two orbiting electrons. An alpha particle is a nucleus of two protons and two neutrons emitted by an unstable larger nucleus during radioactive decay.

Define

Session 6
a) Human sex hormones are used to interfere with the process of egg maturation.

b) Non-hormonal methods prevent sperm from meeting the egg.

Session 7
a) A pathogen is a microorganism such as a virus or bacterium that causes infectious diseases in animals or plants.

b) Mosquitos spread malaria when they bite people. Removing the water would remove the breeding ground so the mosquitos cannot breed. This would reduce the numbers of mosquitos and therefore the number of mosquito bites that spread malaria.

c) Use of a mosquito net when sleeping to prevent mosquito bites.

Session 14
a) During a physical change a substance changes its physical state (solid, liquid, gas, or into an aqueous solution). The substance recovers its original properties if the change is reversed.

b) During a chemical change a chemical reaction occurs where bonds are made and broken, and new substances are made.

Answers to additional questions

Session 27

a) Resistance is the property of a component or wire that opposes the flow of current through it.

b)

	increases the resistance	decreases the resistance
Increasing the length of a metal wire …	✓	
Increasing the temperature of a thermistor …		✓
Increasing the light intensity on a light dependent resistor (LDR) …		✓

Session 29

The nuclei of isotopes of the same element have the same number of protons but a different number of neutrons.

Describe

Session 2

a) Water travels from the cells in the gut to the blood by osmosis. Water travels from a dilute solution to a concentrated solution across the partially permeable membranes of the cells.

b) Water moves into animal cells by osmosis. Cells swell. As there is no cell wall, animal cells will burst if there is an extreme difference between the water content inside the cell compared to outside the cell.

Session 3

The following points describe adaptations of the lungs to absorb oxygen:

- Large surface area due to the shape of the alveoli (air sacs) allows lots of diffusion of oxygen into the blood.
- The lungs, particularly the alveoli, have a good blood supply, which maintains the concentration gradient.
- Moist surface for oxygen to dissolve, which makes diffusion easier.
- Muscles around the lungs ventilate the lungs, which maintains the concentration gradient.
- Short distance – the oxygen only needs to travel a short distance from the air in the alveoli to the blood.

Session 8

a) Rainforests are a habitat for a large number of organisms. Removing this environment removes the habitat, food and mates of organisms so they cannot survive or breed, so there are fewer organisms of different types.

b) Burning fossil fuels for transport produces carbon dioxide gas, which is a greenhouse gas. An increase in production of greenhouse gases leads to an increase in the greenhouse effect. This leads to a rise in the surface temperature of the Earth, causing climate change.

Session 13

a)

Property of Group 1 element	Correct (✓)	Incorrect (✗)
All are metals.	✓	
All are shiny when cut with a knife.	✓	
All react with water producing carbon dioxide gas.		✗
All react with water producing an alkaline solution.	✓	

b) i) The radius of the atoms gets bigger.

ii) The elements become more reactive as you go down the group.

iii) As you go down the group the radius of the atoms of the Group 1 metals gets bigger. This means that the (outer) electron involved in reactions is held less strongly, so is more easily removed. This means that the elements get more reactive.

Session 26

a) i) Energy is transferred electrically from the mains to the light bulb. Some of this is stored as thermal energy in the bulb and some is transferred by light radiation to the surroundings.

ii) Energy is transferred electrically from the mains to the fridge. The fridge uses this energy to transfer thermal energy stored in the food to a store of thermal energy in the air outside the fridge.

b) i) The solid ice particles have less stored kinetic energy than the liquid water particles.

ii) Similarities: particles in both states touch each other; the particles are the same type (water). Differences: liquid particles move from place to place, whereas solid particles only vibrate about a fixed point.

Design

Session 11

Equipment needed: 30 cm ruler OR metre rule; test subject.

Independent variable: Intake of caffeine or not.

Control variables: Same distance of ruler above the test subject's fingers; same test subject; same ruler / metre rule; same method of measuring reactions; same experimental method; same number of repeats; same experimenter.

Dependent variable: the distance the ruler has fallen before the test subject catches it in their fingers. This is measured from the zero to the number just above the test subject's thumb. The experiment should be repeated 10 times before consuming caffeine and 10 times after consuming caffeine.

Risks: Dropping a ruler onto the test subject's foot and causing a bruise / broken toe. The control measure is to ensure that the test subject's feet are well away from the ruler drop zone.

Method: 1. One person holds out their hand with a small gap between thumb and fingers. The second person holds a ruler or metre rule with the zero at the top of the thumb of the first person.

2. The second person drops the ruler without telling the other person. The first person catches the ruler / metre rule.

3. The number level with the top of the first person's thumb is recorded in a table. Repeat 10 times.

4. The first person then drinks a caffeinated drink. Repeat the process and compare the attempts with caffeine to the attempts without caffeine.

Analysing the results: Both data sets should be checked for any anomalies, and these should be ignored; a mean average of each data set should be calculated and the means should then be compared. A conversion table can then be used to convert the mean distances to reaction times. The smaller the mean distance, the faster the reaction time.

Session 22

Equipment needed: Beaker; pencil; chromatography paper; three test pens; wooden splint; water

Independent variable: The type of pen.

Control variables: Same sized dot for each pen; same distance of dot from the surface of the water.

Dependent variable: The number of different colours observed in each pen's chromatography trace. This is observed by eye. The experiment should be repeated once.

Risks: Glassware should be checked for sharp edges before use to ensure that it cannot cut the skin.

Method: 1. Use a pencil to draw a horizontal line 2 cm from the edge of a piece of chromatography paper.

2. Draw a dot on the line in each colour of pen and label each dot in pencil.

3. Put a little water into a beaker to a depth of around 1 cm.

4. Suspend the paper on a pencil or wooden spill and balance on the edges of the beaker so the very edge of the paper touches the water in the beaker.

5. Wait for the water to travel to near the top of the paper and remove the paper. Draw a line to show where the water travelled up to. Let the paper dry.

6. Count how many colours are in each pen's ink, by counting the number of dots of separate colours, and conclude which has the most.

7. Repeat the experiment once to check for any anomalies.

Analysing the results: Check, observe and record the number of colours in each pen trace for each repeat, to see that they are similar.

Determine

Session 10

a)

		Mother (ff) gametes	
		f	f
Father (Ff) gametes	F	Ff unaffected (carrier)	Ff unaffected (carrier)
	f	ff cystic fibrosis	ff cystic fibrosis

b) 50% or 0.5.

Session 12

a) The lower the molecular mass, the lower the melting point and boiling point.

b) i) Naphthalene.

ii) Water.

iii) Hydrogen.

Session 13

a) i) 18

ii) 16

iii) 7

iv) 8

b) Sn

Session 29

Starting activity = 2000 counts/min; half of 2000 counts/min = 1000 counts/min.

This occurs when time = 8 minutes.

Draw

Session 15

a) Sodium atoms donate an electron from their outer shell, forming an ion with a +1 charge. Chlorine atoms accept the electron, completing their outer shell, forming an ion with a −1 charge. The positively and negatively charged ions attract each other.

b)

Session 15

Session 22

Separation techniques

crystallising dish

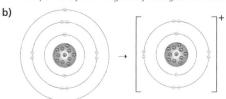

crystals

Diagram should be as shown but could include a source of heat such as a Bunsen.

Session 22

a) Drawing to show / labelled with: conical flask, measuring cylinder, sodium thiosulfate solution, dilute hydrochloric acid, paper, pen, water baths at different temperatures, thermometer, stopwatch, safety goggles.

b) At a higher temperature, the particles have more kinetic energy so more particles have the energy to react when they collide, and the particles collide more frequently.

Session 27

a) i)

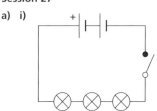

ii)

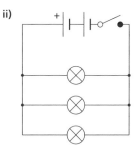

b) The series circuit can have the switch (X) on any wire; the parallel circuit must have the switch (X) next to the battery, on either side.

Session 30

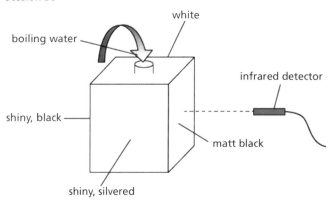

Estimate

Session 23

a) 500 000 N

b) A value more than 500 000 N.

c) A value less than 500 000 N.

Session 25

a) Twice the frequency of the red wave, so 8 Hz.

b) i) and ii)

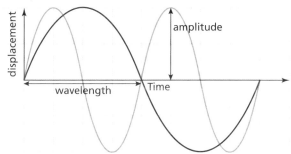

Session 30

As force = mass × acceleration, if the force doubles, then so does the acceleration, so the acceleration will be 4 m/s².

Session 30

Wire length is four times that of the wire with 5 Ω, so resistance is 20 Ω

Answers to additional questions

Evaluate

Session 4
a) i)   Men have smoked since before 1900 and the number of men who smoke increased until the 1960s when the numbers started to level off.

   ii)   Women started smoking in the 1920s and the number of women who smoke has steadily increased to the 1990s. There is no levelling off.

b) i)   Evidence for: the incidence of cancer increases as the incidence of smoking increases for both men and women. Evidence against: even though many men smoked the early 1900s there was not a very high incidence of cancer at this time. Overall, the evidence for the argument that smoking causes cancer is stronger than the evidence that there is not a link.

   ii)   Tar in cigarette smoke coats the inside of the lungs. Tar from cigarette smoke is carcinogenic and can cause the cells on the inside of the lungs to divide uncontrollably. The uncontrolled growth of cells can invade other tissues and result in more uncontrolled growths of cells, also called tumours.

Session 12
a) There is no space between particles in a liquid and in a gas the particles are much further away from each other than suggested here. The size and shape of particles are not realistic.

b) Similarities: The particles of the same substance in the solid state and the gaseous state are the same size and shape. Differences: The same volume of a gas has far fewer particles (so less mass) than the same volume of a solid. The particles in a gas are much further apart than in the solid.

Session 16
a) An electrical current is a flow of charged particles. Both substances have negatively charged particles that are free to move.

b) Both are suitable because they can conduct electricity. Graphite is not suitable because it is brittle and cannot be drawn into a wire, whereas copper can be drawn into a flexible wire. Overall copper is more suitable.

Session 19
a) Aluminium is more reactive than carbon so carbon will not displace aluminium from its compound.

b) It has to conduct electricity, which requires charged particles to be free to move. The charged particles in the mixture are only free to move when the mixture is molten.

c) Both produce a valuable product. Recycling produces a less pure product but is cheaper than extracting aluminium from its ore by electrolysis. Electrolysis uses a lot of energy to produce the electricity. If this is generated from fossil fuels more greenhouse gases will be produced than during recycling. Also, aluminium ore is a finite resource.

Session 20
a) i)   Any pH value between 4 and 7.

   ii)   7

b) i)   Strengths: Limestone chips are made from calcium carbonate. Weakness: Soil is not made from hydrochloric acid; hydrochloric acid has a pH lower than 4.

   ii)   Fizzing, change in pH, which could be seen if an indicator was added.

   iii) and iv) The mass reading will reduce because the reaction produces carbon dioxide which escapes.

   v)   Any two from:
- Increase the concentration of acid.
- Make the surface area of the limestone bigger by breaking it into smaller pieces.
- Increase the temperature.
- Use a catalyst.

Explain

Session 1
a) Ribosomes are too small to be seen under the light microscope.

b) The function of the pancreas cell is to make proteins. The pancreas cells have lots of ribosomes which make proteins. The cell is adapted to its function of making proteins by having lots of ribosomes that make proteins.

Session 3
a) Large surface area due to the shape of the spongy mesophyll cells allows lots of diffusion of carbon dioxide into the palisade cells. Moist surface for carbon dioxide to dissolve, which makes diffusion easier. Stomata that open to allow gases in and out. The air spaces allow for movement of CO_2 from stomata to the palisade cells. Short distance means that the carbon dioxide only needs to travel a short distance from the underside of the leaf where it enters the palisade cells, where it is used.

b) i)   In the chloroplasts.

   ii)   water + carbon dioxide $\xrightarrow{\text{light}}$ glucose + oxygen

Session 7
a) The part of the virus contains a foreign protein that triggers the immune system. The immune system will make white blood cells to fight off the part of the virus and destroy it. Some of the white blood cells that match the virus remain in the body, so if the real virus enters the body the immune system can react quickly to destroy the virus before it can cause the disease.

b) only part of the virus used it cannot cause the disease.

Session 20
A more reactive metal can displace a less reactive metal from its compound.

Magnesium is more reactive than iron so can displace the iron out of the solution of iron(II) sulfate, making solid iron and a solution of magnesium sulfate.

Session 24

a)

Factor	Thinking distance	Braking distance	Both
Speed of car			✓
Water on road		✓	
Driver's tiredness	✓		
Driver's alcohol consumption	✓		
Condition of car's brakes		✓	

b) A fully loaded van has more mass, so when travelling at the same speed (as an unloaded van), it has more kinetic energy. This means that the brakes must do more work to stop a loaded van. The brakes produce the same amount of braking force when the van is loaded and unloaded. This means that when travelling at the same speed, a loaded van takes a longer distance to stop than an unloaded van.

Justify

Session 14
a) i)   Physical change was the students breaking up the calcium carbonate into smaller pieces. Heating in the flame made a chemical reaction happen so there was a chemical change.

   ii)   We know there was a chemical change because there was a change in mass, so some atoms were lost as a gas.

b) i)   calcium carbonate $\xrightarrow{\text{heat}}$ calcium oxide + carbon dioxide

   ii)

Atom	Number in $CaCO_3$
Ca	1
C	1
O	3

Session 16
a) Solid.

b) Substances need free charged particles to move in order to conduct electricity. In the solid state the ions are fixed and unable to move. In the liquid state the charged particles are free to move.

Session 17
a) i)   A and B.

   ii)   C.

b) A and B have only one type of atom, whereas C has more than one type of atom joined by a chemical bond.

Session 21
a) Change in colour.

b) The temperature of the water rose by 9 °C, indicating that thermal energy has been released to the surroundings from the reaction. This means that the reaction must be exothermic.

Session 29

The diagram shows that gamma radiation can penetrate paper and aluminium, but not lead. Gamma radiation is the most penetrating of the three different types of nuclear radiation. Lead is a very dense substance and can absorb gamma rays. Gamma radiation sources are kept in lead-lined containers so that the radiation cannot pass through the container.

Session 30

The lamp makes the waves clearer on the viewing screen so they can be measured more accurately.

Plan

Session 11

Equipment: Test tubes, test-tube rack, iodine, starch solution, amylase solution, a range of pH buffers, pipettes, dropping tile, stopwatch.

Diagram:

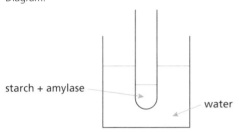

starch + amylase → water

Independent variable: the pH of the solution.

Control variables: Same enzyme; same enzyme solution concentration; same starch; same starch solution concentration; same iodine solution concentration.

Dependent variable: The time for the solution to change colour from blue/black to brown. Measured using a stopwatch.

Repeat method three times, identify any anomalies, remove any anomalies from the dataset, calculate the mean time for the colour change for each pH value tested. Plot a graph of mean colour change time (*y*-axis) against pH (*x*-axis).

I expect there to be an optimum pH where the time for the colour change is shortest, and the times will be higher for lower and higher pH values.

Method: 1. Set up a range of solutions with different pHs (within a range of pH 4 to pH 9) using buffer solutions in test tubes.

2. Set up a dropping tile with iodine in each well for each pH.

3. Add a mixture of starch and amylase to each test tube and start a stopwatch.

4. Remove a sample of the solution from each test tube every (suitable time period; e.g. 20 seconds).

5. Record how long it takes for the blue/black of the iodine to change to brown.

6. Repeat three times.

Safety: Iodine will stain so take care not to spill it on clothes or skin. Iodine can be an irritant, so wear safety goggles. Glass coverslip and slide can break easily – handle carefully. Risk of slip / fall due to spillages of liquids – clean up spills straight away.

Session 22

Equipment: Metal blocks with two holes in each; thermometer; petroleum jelly; 50 W, 12 V heater; 12 V power supply; insulating material; voltmeter; ammeter; connecting wires; stopwatch; electronic balance; safety goggles.

Independent variable: The material of the block.

Dependent variable: The temperature change of the block.

Control variables: The start temperature of the block; the power of the heater; the heating time interval; the insulation system (material / thickness / area); the ammeter and voltmeter values.

Analysis: Calculate the temperature change for each experiment using:

$$\text{temperature change} = \text{final temperature} - \text{initial temperature}$$

Calculate the mean temperature difference for each metal block.

3. Calculate the actual power output of the heater using measured values of potential difference and current:

$$\text{power} = \text{potential difference} \times \text{current}$$

4. Calculate the energy transferred using:

$$\text{energy transferred} = \text{power} \times \text{time (in seconds)}$$

5. Calculate the specific heat capacity of each block using:

$$\text{specific heat capacity} = \frac{\text{energy transferred}}{\text{mass} \times \text{mean temperature change}}$$

6. Compare the specific heat capacities for each metal.

Method

1. Measure and record the mass of each block.

2. Wrap an insulator around the block.

3. Smear petroleum jelly around the thermometer and put the thermometer in the hole in the metal block. Measure and record the initial temperature of the block.

4. Connect the heater to a circuit with a voltmeter and an ammeter and connect to a power supply.

5. Put the heater in the second hole in the metal block.

6. Switch on the heater, and monitor the ammeter and the voltmeter to check that the current and potential difference stay steady throughout heating. Measure and record the voltmeter and ammeter readings.

7. After 10 minutes, turn off the power supply.

8. Measure and record the maximum temperature of the metal block, and then calculate the temperature change.

9. Calculate the power of the heater and the energy transferred using the potential difference and current readings.

10. Repeat for each metal block and compare.

Safety: Eye protection; allow the hot metal blocks to cool before handling them, to prevent burns.

Plot

Session 24

Graph should have:

• *x*-axis labelled with time in seconds, with a suitable scale
• *y*-axis labelled with velocity (m/s), with a suitable scale
• accurate plotting and suitable line drawn.

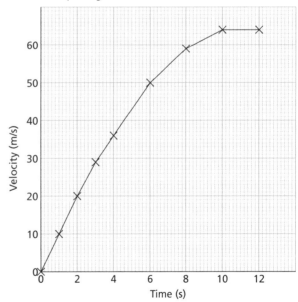

Session 24

a) $\text{acceleration} = \dfrac{\text{change in velocity}}{\text{time taken}}$

   i) $\dfrac{13 - 0}{5} = 2.6\,\text{m/s}^2$.

   ii) $\dfrac{30 - 13}{2} = 8.5\,\text{m/s}^2$.

Answers to additional questions

b)

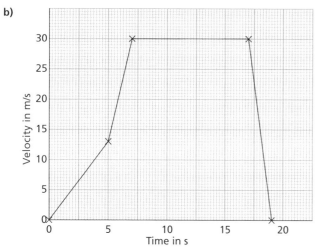

Graph should have:

- *x*-axis labelled with time in seconds, with a suitable scale
- *y*-axis labelled with velocity (m/s), with a suitable scale
- accurate plotting and suitable line drawn.

Predict

Session 11

a) **i)** Water will move out of the cells and the beetroot will have less mass.

ii) Water will move into the cells and the beetroot will have more mass.

iii) Water will not move into or out of the beetroot cells so the mass of the beetroot will stay the same.

b) There would be a net movement of water out of the beetroot cells into the concentrated sugar solution down its concentration gradient. Pure water contains a higher concentration of water molecules so there will be a net movement of water into the beetroot cells from pure water. A sugar solution with the same concentration will have no overall effect on the mass of the beetroot cells as there will be no net movement of water.

Session 11

Food test	Outcome of the test (positive = ✓ / negative = ✗)
Benedict's test for sugars	✗
Iodine test for starch	✗
Biuret test for protein	✓
Ethanol test for lipids	✓

Session 12

a) The density of other substances in the solid state is more than the density of the substance in the liquid state.

b) **i)** As ice melts it becomes a liquid, so sea levels rise.

ii) Any two from:

- flooding of coastal areas
- contamination of fresh water with salt water
- removal of ice habitats.

Session 28

a) 24 paper clips can be supported.

b) A number bigger than 12 can be supported.

Show

Session 5

Reflex actions are a response to a change that can cause harm, such as a hot surface, dust in the air or falling. The change in the environment is detected by a receptor. This results in an impulse in a sensory neurone that connects to a relay neurone. A relay neurone is used because it is faster than the brain as no thinking is involved. The relay neurone connects the impulse to a motor neurone, which connects to an effector, such as a muscle, which brings about a response. The whole action is fast so harm is prevented.

Session 9

a) **i)** Testes.

ii) Ovaries.

b) Egg cells and sperm cells have a random selection of half of the DNA of a normal body cell.

c) **i)**

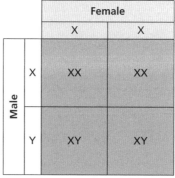

ii) Female ovary cells produce eggs with only the X chromosome. Male testes cells produce sperm containing either an X chromosome or a Y chromosome. Two of the possible egg-sperm combinations are XX – females, and two are XY – males, so there is a 50% chance of a male or female embryo being formed.

Session 22

Measure and record the length, width, and height of the cube with a ruler.

Place the cube on an electronic balance and measure and record its mass.

Calculate the volume of the cube using the equation:

$$\text{volume} = \text{length} \times \text{width} \times \text{height}$$

Calculate the density of the copper using the equation:

$$\text{density} = \frac{\text{mass}}{\text{volume}}$$

Session 22

Place the coal on a balance and measure and record its mass.

Fill a displacement can with water up to the bottom of the spout.

Place a measuring cylinder under the spout to collect the water that is displaced.

Place the coal into the can and collect the water in the measuring cylinder.

Measure and record the volume of the water.

Calculate the density of the coal using the equation:

$$\text{density} = \frac{\text{mass}}{\text{volume}}$$

Session 25

a) From left to right: gamma rays, ultraviolet, microwaves.

b) **i)** Medical imaging.

ii) Sight.

iii) Remote controls / fibre optic communications.

iv) Communications.

Sketch

Session 11

See key ideas card B11.4. Sketched diagrams of:

a) Adding iodine to a food.

b) Heating a food sample with Benedict's solution.

c) Adding Biuret's reagent to a food sample in solution, either as one solution or as Biuret A and B in succession.

d) Adding Sudan III to a food sample mixed with water.

Session 22

a)

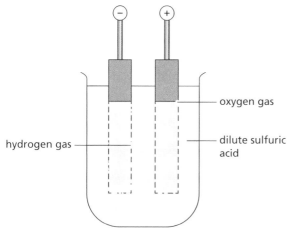

Diagram should show the cathode and anode in a container, immersed in a solution of sulfuric acid and connected to a power supply in a circuit, which can be suggested by '+' and '−' symbols.

b) A lit splint inserted into the sample of gas ignites quickly with a pop sound in the presence of hydrogen.

Session 22

a) Water is in short supply in some countries. It is expensive to remove salt from sea water to make water fit for drinking. Water is necessary for life (for chemical reactions and cooling in the human body). Every human should have the right to clean water. Conclusion that the expense is justified.

b) Diagram as shown in *key idea card C17.8*.

Session 30

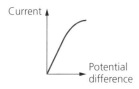

Suggest

Session 1

The function of the palisade cell is to do photosynthesis. The palisade cells have lots of chloroplasts which do photosynthesis. The cell is adapted to its function of photosynthesis by having lots of chloroplasts that use sunlight to do photosynthesis.

Session 8

a) i) Biotic: availability of food.

Abiotic: any two from:
- hot
- little water
- windy (sand storms)
- sandy and unstable ground.

ii) Any two from:
- feet have a large surface are to prevent sinking in ground
- does not sweat to prevent water loss
- thin fur to prevent overheating
- long eyelashes to protect the eyes from sand
- hump of fat to store nutrients and water as fat.

b) Any two from:
- thick waxy cuticle
- small leaves or spines
- hairy leaves
- leaves curl
- deep roots.

Session 11

a) Mark–release–recapture.

b) Use a quadrat placed at random.

c) Use a transect line (with a quadrat).

Session 18

a) Longer molecules have stronger forces between their molecules so more heat is needed to overcome them.

b) Fractional distillation takes place in a tall tower. The temperature of the tower decreases as you go up the tower. At different points on the tower, liquids or gases can be piped off. The crude oil is heated at the bottom, the mixture boils and mixture of gases is formed. The gases rise up the tower. As each gas reaches the surface that is at the same temperature as its boiling point the gas condenses and the liquid is run off. The other gases continue upwards until they reach the surface at their boiling point.

c) Cracking

Session 21

a) The thermal energy gives the reaction its activation energy.

b) Burning (combustion) requires oxygen. Reactions happen when particles collide frequently with enough energy for a reaction to happen. If there are more oxygen particles with enough energy then collisions with the magnesium particles will happen more often, so a reaction is more likely, so it will react at a faster rate.

Session 21

Any three from:
- increase temperature
- use a catalyst
- increase surface area
- increase concentration (increase pressure).